C000245119

THE OVERPRODUCTION OF TRUTH

THE OVERPRODUCTION OF TRUTH

Passion, Competition, and Integrity in Modern Science

Gianfranco Pacchioni

Great Clarendon Street, Oxford, OX2 6DP,
United Kingdom

Oxford University Press is a department of the University of Oxford.
It furthers the University's objective of excellence in research, scholarship,
and education by publishing worldwide. Oxford is a registered trade mark of
Oxford University Press in the UK and in certain other countries

An earlier version of this book has appeared as *Scienza, quo vadis? Tra passione intellettuale e mercato* © 2017 Società editrice il Mulino, Bologna

First Edition published in 2018

Impression: 1

Published in the United States of America by Oxford University press
198 Madison Avenue, New York, NY 10016, United States of America

British Library Cataloguing in Publication Data
Data available

Library of Congress Control Number: 2018940066

ISBN 978–0–19–879988–7

DOI: 10.1093/oso/9780198799887.001.0001

Printed and bound by
CPI Group (UK) Ltd, Croydon, CR0 4YY

Contents

Acknowledgements

This book follows many chats and discussions I have had in recent years, often in front of a glass of beer or wine, with colleagues of my generation. More and more often, the conversation slipped onto the topic of how science is changing around us. However, I wouldn't have gotten down to write a book on this subject had it not been for my daughter Giulia, with whom I began to discuss these issues at the time of her PhD in physics at the Ecole Polytechnique Fédérale in Lausanne. Through her experiences and those of her young colleagues, I discovered the other side of the coin and tried to look at science through the eyes of a student getting into the subject in the new millennium. With Giulia I could talk about things that for various reasons have been difficult to discuss with my own students and young collaborators. Over the years she pointed me towards a number of articles, blogs, discussions, curiosities, and anecdotes on the subject. This allowed me to form a less personal and more general idea of the dimensions of the problem, and to transform into a solid perception what would have otherwise been nothing more than superficial feelings. No surprise, then, that Giulia was the first reader of the manuscript and made the first remarks on it.

Of course, many valuable suggestions came also from colleagues (of my generation, I confess). In particular, I am grateful to Elio Giamello and Dario Narducci for their advice and comments (if I did not welcome them all, it was just not to transform an agile pamphlet into a heavy textbook).

The English version would simply not exist without the stimulus, support, and continuous help of Angelo Gavezzotti, who after reading the book helped me to overcome the barrier of the translation. He made an impressive number of improvements in the English version as well as valid comments on the text, and I really owe him.

Finally, there is another person who has played an important role throughout the story and helped to define how this book is structured. I'm talking about Alessia Graziano, my editor at the Italian publisher *Il Mulino*. The first draft of the book had all the typical features of an academic text: erudite, referenced, detailed, rich in tables and charts. In short, tedious and boring, almost unreadable. She advised me to come at it from a more personal angle—lived, participated, and therefore more enjoyable. That's what I tried to do. If I did not succeed, the demerits are all mine.

Introduction

About 10,000 years ago, at the beginning of the agricultural revolution, on the whole earth lived between 5 and 8 million hunter-gatherers, all belonging to the *Homo sapiens* species. Five thousand years later, freed from the primary needs for survival, some belonging to that species enjoyed the privilege of devoting themselves to philosophical speculation and the search for transcendental truths. It was only in the past two hundred years, however, with the advent of the Industrial Revolution, that reaping nature's secrets and answering fundamental questions posed by the Universe have become for many full-time activities, on the way to becoming a real profession. Today the number of scientists across the globe has reached and exceeded 10 million, that is, more than the whole human race 10,000 years ago. If growth continues at the current rate, in 2050 we will have 35 million people committed full-time to scientific research. With what consequences, it remains to be understood. For almost forty years I myself have been concerned with science in a continuing, direct, and passionate way. Today I perceive, along with many colleagues, especially of my generation, that things are evolving and have changed deeply, in ways unimaginable until a few years ago and, in some respects, not without danger. What has happened in the world of science in recent decades is more than likely a mirror of a similar and equally radical transformation taking place in modern society, particularly with the advent

The Overproduction of Truth. Gianfranco Pacchioni.

DOI: 10.1093/oso/9780198799887.001.0001

of new forms of communication, the Internet being the most prominent.

In 1906, the Danish philologist Johan Ludvig Heiberg found in the ancient monastery of the Holy Sepulcher in Constantinople a letter from Archimedes, addressed to Eratosthenes of Alexandria, in which some geometric theorems are shown:

> I am sending you some theorems I had discovered by limiting myself to give you the statements and inviting you to find the proofs that I had not yet indicated [. . .] of those theorems for which Eudoxos found the first proof about the cone and the pyramid, that is, the cone is the third part of the cylinder and the pyramid is the third part of the prism having the same base and height. For this no small part [of merit] must be attributed to Democritus, who first made this property known.

We are in 200 BC but the language and the kind of discussion display a surprising modernity. Archimedes quotes Eudoxos (408–355 BC) and Democritus (460–370 BC), who provided contributions in the field of geometry centuries before him, precisely as we do today when we include bibliographic references in a scientific work.

Not so different is the way in which scientific controversies developed in the past. Take the rather harsh one that arose at the beginning of the eighteenth century between Newton and Leibniz about who had invented infinitesimal calculus. In a letter to Abate Conti in December 1715, Leibniz so expressed himself: 'I come immediately to the question that concerns us. [. . .] There is no evidence that Newton had discovered before me the characteristic and the algorithm of infinitesimal calculus, though it would have been easy for him to discover them, had he thought about it.'[1] Abate Conti responded three months later in March 1716:

> I was so late in answering you because I wanted to join my letter with Newton's response to your comment [. . .]. I came to the conclusion that the question comes down to establishing whether

> Newton has found infinitesimal calculus before you, or if you have found it before him. You were the one who published it first, that's true; but you have admitted that Newton has let you have a look at some aspects in some letters.

We can say that the way of doing science until the end of the past century did not depart much from these models of discussion and types of confrontation: personal, based on direct acquaintance, and on mutual respect. I still keep some letters exchanged with colleagues in the 1980s that recall closely, proper proportions made, those mentioned above. From when I started my scientific activities, and for at least the first ten to fifteen years, how research was conducted was not much different from the methods used at the times of Fermi, Pasteur, and even Volta. Of course, air transport had reduced travel times, the spread of the telephone had facilitated contacts, but in the end the way to proceed and generate new knowledge was not largely different. The pacing was such as to allow—besides the creation of new ideas and testing them through original experiments—plenty of time to think about how and why things are. This path was at the root of our civilization and our culture. Then the Internet came on the scene, and in a short time everything has changed. How we communicate, share results, debate within the community—everything has been overwhelmed by the procedures and tools afforded by the digital revolution. There have been great positive implications, such as making available immense amounts of information previously concealed in remote libraries, but there have been some negative aspects that have quickly evolved into practices, behaviour, and mental attitudes that contradict the ethical principles that have supported the development of modern science. For those like me who have had the privilege (or misfortune, depending of your point of view) of experiencing both environments, the contrast is sharp and evident.

Today the world of scientific research dwells in a time that twists the sense of things, a continuing and spasmodic race to publication, to the production of results of little or no issue that are not necessarily aimed at increasing our knowledge; it is a projection always and exclusively to the future with a worrying tendency to dismiss the past as if it had never existed. In order to emerge, young people must accept to be involved in a tight and ruthless competition that leaves no room for meditation, originality, or risk, all factors that should be part of every proper research activity. There is not much time left to tackle unusually complex themes, or for the thrust of investing energy in projects with little chance of success. The pressure toward achieving new results is daily, and leaves little room, if any at all, to ponder the meaning of what one does. In this way, in a short time the world of research has changed from the passionate activity of a few selected people to a crowded universe of practitioners, often with few ideas and sharing little or no ethical values. It is too early to predict how and how much this will affect the development of scientific thinking and the relationship between science and society in the future. In my opinion there is little doubt that this will induce changes, but not necessarily good ones.

What follows is not a scientific treaty and does not want to be one. It is the story of an individual course, as seen through the eyes of someone who has experienced the changes that have occurred in the world of science over the past forty years. It is a collection of experiences, anecdotes, actual stories, personal reflections, all accompanied by objective data and findings, as well as by quantitative analyses that support and strengthen what until recently were just sensations for me. In the past few years I have collected, read, and studied reports, scientific papers, documents, surveys, articles on the specialized press, blog discussions, all about trying to understand the trends sweeping the science arena today and to some extent shaking its foundations. From the in-depth and meditative analysis of this reality stems the question that is behind the title to this book: Science, where are we going? Oscar Wilde said that questions are never indiscreet. Sometimes, the answers are. Paraphrasing, we can say that the question of where science is going may not be embarrassing. The answer could be, though.

1

No progress without basic research

Figure 1 shows an object that may be recognized by only a few intimates. Yet, it played an important role for many of us (and surely for me!) for a certain period of the past century. It is the rotating head (typeball) of an IBM typewriter. I used it in the early 1980s in Berlin to write my PhD thesis. Precisely, it was in late 1983 when I was facing the challenge of writing my work. The very first personal computers had appeared, but they were still very primitive, almost unknown, and no one possessed sufficiently developed software for text processing.

After some analysis and consultations I decided to entrust myself to the old, dear, and faithful typewriter. I had brought from Milan the mythical Olivetti 'Lettera 22', a mechanical typewriter that seldom missed a shot. But, unfortunately, it was too slow, and the hammers were heavy to press. In short, it was a typewriter, but nothing more. Sometime before, I had devoted time (with great effort) to a course for typists, to learn ten-finger typing without looking at the keyboard. The evening course had its charm as I was the only male in a sea of female would-be secretaries, many of whom were young and pretty. But it was my girlfriend who had convinced me to accompany her to the course, which greatly reduced the potential for interaction with the rest of the 'class'. Even though I later did many other difficult things in my life, I must recognize that writing a text by reading it from left to right using all ten fingers, and in a hell of a racket caused by 30 simultaneous typists, remains one of the most complex

The Overproduction of Truth. Gianfranco Pacchioni.

DOI: 10.1093/oso/9780198799887.001.0001

Figure 1 A typeball from an IBM electric typewriter of the late twentieth century.

Source: Adapted from data in Hudson et al. (2011), published in *Proc R. Soc London*, vol. **278**.

exercises that I have ever experienced. Eventually, however, I made it, I had my nice diploma as a typist, and I wanted to use it. But in order to see my knowledge be fruitful, I needed an electric typewriter. This was not difficult: all that was needed was to buy my boss's secretary a couple of coffees to convince her to lend me an electric IBM with rotating head that she did not use and kept 'in reserve'. It was an extraordinary piece of jewelry compared to my Olivetti, with an impressive writing speed just thanks to the bouncing head that went up and down, automatically selecting the characters with very fast rotations, and imprinting on the paper all symbols with even thickness and well aligned. At that point in time, it was a miracle of technology compared to a traditional, manual typewriter. Since I wrote my thesis in the evening and over weekends, I used to take home the typewriter. The mythical IBM was red, incredibly heavy, and occupied entirely the (small) dining table of the tiny home we lived in. I thrived upon it and produced my precious document, the result of three intense years of work, prepared by pasting the figures, creating the tables by hand, correcting the unavoidable mistakes in the text with the appropriate ink (there were very few options for rewriting or reformulating text, short of typing from scratch a whole new page if not the entire chapter).

Today, as I write these pages, I sit comfortably in my armchair, with the lightweight processing power tool that is my laptop on my knees. I do not care about mistakes, because there is an automatic debugger, nor about the layout that just deploys as I write. I don't even have to collect the pages of my text, as everything remains 'written' in a magnetic memory. My words don't have to be physically laid out in indelible marks on sheets of paper. If I wanted to, I could just re-read everything on the screen and send it to the publisher. Perhaps unfortunately, because I belong to a generation that was educated on books, I can hardly get used to the idea of just reading via an electronic support, so I know I'll have to print, hold in my hand, and smell the paper, to make sure everything really exists.

Science and technology change our way of life

This simple example helps us understand how deep and incisive the changes in our everyday lives have been in the short span of a few decades. The transformations around us have been epochal, probably the deepest and fastest humankind has witnessed in its history. I am not overstating. Every now and then I talk to groups of boys in secondary schools. They are 17–18 years old, and

I show them a set of everyday, commonly used objects or devices: a smartphone, a tablet, a webcam, and intangible but popular tools such as Facebook, Twitter, and Skype. At that point I tell them an uncomfortable, but simple and disruptive truth: when you were born all this did not exist. With this I shatter their implicit belief that the world in which they live has always existed the way they know it. And many of the things we use every day and which are indispensable for us came into common use only a few years ago. Because these changes, even if almost imperceptible, keep contributing to our existence, the next question I ask the boys is: 'If these things were not here 15–20 years ago, what should we expect in the next 15 or 20 years? In short, how will our future be?' But anticipating the future is impossible, I tell them, except for a few categories of superior people, such as financial analysts and astrologists (with what rate of success, it is for you to decide).

Imagine if at the beginning of the 1990s someone had told a person of my generation that in twenty years practically every person on the face of the Earth will have in his or her pocket a small object weighing a few tens of grams that will allow us not only to communicate in real time with everyone else on the planet, but also to access all sorts of information, watch movies, listen to music, send documents, pay the dentist, photograph your grandmother at the restaurant, send the picture to a cousin, expose your emotions, tell personal facts to all those who want to know them, and so on. That prophet would have been taken for crazy (and maybe even dangerously crazy). But all this is now a fact. And it is so thanks to technology, which is nothing more than the translation into practice of the advances in knowledge connected with scientific research, with the discovery of new phenomena and the definition of new theoretical models. These are all things that lead to practical applications and to direct consequences that, like it or not, deeply change our lives and our way of being.

This is science. Mankind has always used it to try to answer fundamental questions that began emerging as soon as our species was released from the daily struggle of survival in a hostile world. When the turnover from hunter-gatherer to farmer took place, as soon as the knowledge needed to cultivate land and raise livestock had been developed, man began to face the deeper questions about his own existence and his relationship with nature. And he began to know, to understand, to tear the veil from nature's secrets in a slow but never-ending process that led him to develop civilization, in turn based on the development of technologies.

In this process, some discoveries or inventions more than others have marked the path of mankind. Near the end of the recent millennium, the same question was asked to a large sample of historians, scientists, and journalists: 'Which, in your opinion, is the most important discovery or invention of the last thousand years?' Please consider that in this period of time such achievements as the steam machine, penicillin and antibiotics, DNA's structure, the transistor and the computer, planes, cars, television, and many others were made, not to mention fundamental scientific theories such as universal gravitation, Darwin's theory of evolution, Einstein's relativity, the laws of genetics, and quantum mechanics. The list is really long, and anyone can add more items. Actually, many among the interviewed agreed that the most impressive invention of the past thousand years was that of Johannes Gutenberg, the German printer who in Mainz, in the fifteenth century, introduced the mobile printing press. Gutenberg started a real revolution: the free circulation of ideas and knowledge. Until then, books were handwritten and therefore extremely rare and expensive, and hardly accessible, preserved as they were in monastery libraries. There were no more than 20–30,000 books in the whole world in Gutenberg's time, most of them Bibles. Just fifty years after the introduction of the printing process, over 30,000 different titles had been published for a total of twelve million volumes! The price of books plummeted, and knowledge and information began to spread among increasingly large tiers of the population, with a decisive contribution to the rapid development of philosophical thought, of scientific progress, and of culture.

Everybody knows Gutenberg and his fantastic invention, but only a few are aware that mobile printing is based on two fundamental technological innovations: on the one hand, Gutenberg (an expert metallurgist) was able to prepare lead, antimony, and tin metal alloys to make movable characters that did not deform under the pressure of a screw press; on the other hand, he used paper, a flexible, economical, and steady support for embossing print characters in ink. Invented in China around AD 105, paper arrived in Europe almost a thousand years later. This shows that even radical and revolutionary innovations like Gutenberg's do not start from scratch but rely on previous technologies, materials, and knowledge. Personal and collective social progress is therefore intimately linked to the development of scientific research and to technological advances. A few data are enough to let us understand how this coupling has influenced the history of the previous century. In 1863, the average life expectancy in Italy was below 50; today, it has reached almost 80 years for men and 85 for women. In 2011, again in Italy, there were 2,084 deaths of children under 5 years old; in 1887, that number was 399,505. In just over a century it has gone from 347 to about 4 deaths per thousand births.

Devastating and diffuse diseases such as gastroenteritis, diphtheria, tetanus, and typhoid fever, some of the scourges that affected randomly but in a very painful way families across Europe, practically disappeared through antibiotics, enhanced environmental hygiene, effective vaccination campaigns, and advanced systems for prevention and health care. We tend to overlook the fact that all of this is the result of scientific research. Over the past twenty years, the epochal revolution of the Internet has distorted our way of working and thinking. We are all connected and in touch with one another (which does not relieve the sense of solitude of many of us). We have access to an infinite amount of real-time information without having to consult dusty and heavy volumes stored in uncomfortable libraries. For transportation, we move with a cadence, frequency, and speed that humanity has never known. Yet, quite a few still believe that science and technology are hostile, do harm to human activities, shift us away from our fundamental values, our dimension, or from a balanced relationship with nature. Of course, the damage due to uncontrolled exploitation of the planet is in front of our eyes, and an assessment of what to do in the future is indispensable. But rather than due to science and technology, this has to do with certain models of economic growth in which the everlasting, steady increase of consumption is considered an indispensable prerequisite to social welfare. However, the strive for a return to our origins, for the dismissal of the benefits of technological advancement, as is sometimes advocated in the bucolic yet unrealistic programmes of extreme alternative groups, is by no means easy, nor is it to be taken for granted.

Back to the past?

A few years ago, the BBC featured an interesting reality. Among numerous eligible to attend, producers of a reality TV show chose three families willing to try living for a whole month under the same identical conditions as Edwardian London of the early 1900s. The agreement provided for subsistence and acceptance of living conditions for all members of the family throughout the game, pending exclusion from the programme and waiver of the reward. The first action was to allocate the three families to three different living standards: high bourgeoisie, middle class, and urban proletariat. Not surprisingly, the first were the luckiest, being allowed to take possession of a beautiful two-storey, many-room home, servants and butler included. They had nothing to do with the terrible conditions that the third family was called to face: forced to live in a dingy, cold, and humid environment, subsisting on an insufficient, low-quality diet. This points out a rule that applies to every age: if you are rich, you live better. But be assured that even for our heroes of the high bourgeoisie, the experience turned out to be a rough journey. You should imagine living with little provision of lighting, with some access to the fireplace but a chilling coldness in many rooms, and with no chance of listening to music or, of course, watching TV. This is not to mention the social structure of the family, very much set on a hierarchical relationship between father and children, where a rigid and sometimes brutal etiquette had to be observed. And what about the little daily pleasures we enjoy these days? There was very little or no hot water and a lack of body care products like beauty creams, toothpaste, hair shampoo, and hair dryer. In short, even the typical life of an upper class British bourgeois family turned out to be a tedious, uncomfortable, somewhat boring and, in the end, rather unhappy experience for our 'guinea pigs'.

If this was the perception of the rich family, you can imagine that of the middle class and of the proletarian family. Here it was not just a matter of giving up some items of day-to-day well being so rooted in our habits that we can no longer live without; it was a real struggle for survival. Without refrigerator, washing machine, electric light, and of course service staff, our families' days were consumed in a series of repetitive acts, laborious but essential to ensure everybody's survival, always at a certain level of economic constraint, if not complete poverty. In short, this was an experience that deeply touched all protagonists. And this without having to come to grips with the real and dramatic challenges: to give birth in a hospital in the early 1900s, to get sick with one of those diseases we talked about previously, to fall victim to an accident that could restrict, even temporarily, the ability to work.

After finishing the experience and ending the programme, an attentive onlooker would enormously appreciate many of the conveniences that progress has brought us and that are embedded in our civilization. Living without the benefits of progress is indeed possible, but it is very hard, as were the very hard lives our ancestors faced until recently. For most of us, such a life would simply be intolerable and unacceptable.

Thus, science and technology advances have a clear benefit. They produce important and radical consequences, much to alter (for the better, usually) our existence irreversibly and irrefutably. But this is a complex, slow, non-linear journey, where the light at the end of the tunnel is often invisible, and only the determination and curiosity of researchers have overcome the often-unpredicted catches and difficulties standing in the way. At the heart of these advances is basic research, often called curiosity driven, whose only motivation is to unveil nature's secrets, without necessarily foreseeing a direct advantage, benefit, or utilitarian spinoff. And this aspect is the most difficult to explain to those that do not practice research, but are curious about its modes and rituals, and about the mechanisms that progressively lead to what the public normally calls 'a discovery'. A Spanish colleague, a theoretical chemist like me, recalled his son answering the classic question asked by the teacher, 'What does your dad do?': 'My dad is a scientist . . . but he has discovered nothing so far!' And, in fact, it is true for many of us concerned about science that progress consists of often modest, almost imperceptible advancements, and very seldom of the sudden flash of light that collective imagination so much associates with the idea of discovery.

That flash may happen, but science is patience, study, perseverance, dedication, determination, and so much more.

What is basic research for?

There is a question whose answer is difficult for people dedicated to basic research: 'What is it for?' The story goes that in 1800 Napoleon wanted to meet the famous physicist and chemist Alessandro Volta, whom Napoleon had heard of as a great inventor. Volta took the opportunity to proudly show his pile of copper and zinc discs that could generate a weak electric current. After observing the strange object with interest, Napoleon could not refrain himself from asking, 'What is it for?' Obviously, Volta could not say that a century later electricity would revolutionize the modern world. Much less could he foresee that with highly sophisticated 'batteries', called fuel cells, more than 150 years later man would be able to go to the Moon, nor that, thanks to similar devices, one day ordinary people would walk around with powerful computers in their pockets. Nobody knows what Volta replied, but we know that the question of what is the 'benefit' of a discovery often has no immediate answer, and that much time may pass between the moment when an important phenomenon is discovered and understood and the moment when it becomes of interest and benefit to all. Research must have a general purpose, but needs not be aimed at a precise target, to a practical application to be obtained shortly after. Many great achievements in scientific research have appeared out of the pure curiosity of those who, by observing nature, have drawn conclusions and universal lessons, only later translated into concrete benefits.

To better understand the relationship between fundamental research and technological progress, I will tell the story of a technology we are all familiar with, although most of us are not aware of its origin and of how much intellectual effort and research work has been required to get it working. These days if you need a diagnostic check in a hospital (hopefully not), you may have to undergo a magnetic resonance imaging examination. Everyone in the hospital knows this term, and, I assume, so do many readers. But few know that this term is actually 'truncated' and that the real, complete name is nuclear magnetic resonance, or NMR. Indeed, we are dealing with a technique based on the effects of nuclear physics. So as not to alarm further the already frightened patients who have to enter a huge hollow-shaped magnet, the adjective 'nuclear' was chopped off and things seemed more reassuring. Today there are about 30,000 magnetic resonance devices around the world, and millions of patients are exposed to this risk-free and non-invasive examination every year. But how did we get to that?

The story starts a long time ago.[2,3] Around 1930 Isidor Isaac Rabi (1898–1988) and his group worked at Columbia University in New York trying to measure the magnetic properties of some atomic nuclei such as hydrogen, deuterium, and lithium. Rabi demonstrated that these atomic nuclei can reverse the main direction of their magnetic moment by means of an oscillating external magnetic field, in practice inventing NMR spectroscopy. For his work Rabi was awarded the Nobel Prize in Physics in 1944.

Though Isidor Rabi is generally credited for the discovery of the NMR technique, his work was carried out in a completely artificial context, and therefore far from any possible practical application. Rabi used a molecular jet in high vacuum, where atomic nuclei are far apart and isolated from the surrounding environment. It was a situation devoid of any interest for the public, who deal in everyday life with solids, liquids, and possibly gases but at infinitely higher concentrations than those used by Rabi. One had to wait until 1945 before two independent groups, led respectively by Felix Bloch (1905–83) at Stanford and by Edward Purcell (1912–97) at the Massachusetts Institute of Technology in Boston, showed almost simultaneously the possibility of measuring the magnetic resonance of nuclei in condensed matter, water in the case of Bloch, and paraffin for Purcell. The papers that conveyed these discoveries were published as short Letters to the Editor in January 1946, in the journal *Physical Review*. This achievement and the subsequent work of Bloch and Purcell were awarded the Nobel Prize for Physics in 1952. If someone had asked Rabi, Bloch, and Purcell about what their discovery could be good for, they would hardly have been able to provide a consistent answer.

By 1950, however, it was recognized that the phenomenon of resonance depends not only on the type of nucleus (hydrogen, carbon, etc.) but also on its 'chemical environment'. This means that a hydrogen atom in a molecule behaves differently depending on what atoms are nearby: carbon, oxygen, and nitrogen atoms, for example, all atoms that are present in our body. Suddenly, a secondary appreciation of the original discovery became an important step towards practical use. In fact, analyzing how hydrogen nuclei behave in an unknown molecule can be traced back to what is in the molecule itself, opening the way for the technique to be used to deduce the composition of chemical samples. At that point the usefulness of the NMR technique for chemical analysis became obvious, and interest for it grew at a fast pace. However, important issues remained unsolved, for example its low sensitivity, implying the need to use concentrated solutions. In 1960, Weston Anderson first applied the mathematical technique known as Fourier transform to improve the sensitivity of NMR (Fourier was a French mathematician and engineer at the time of Napoleon and the French Revolution). Again, it was a very important step in the right direction, but not yet leading to the final result.

In the years between 1950 and 1970, the focus was on using NMR for chemical research, and it was only in the late 1960s that the technique began to be applied to biological tissues. Meanwhile, the resonance process of nuclei was supplemented with another component that would then prove very important, the relaxation time. In 1971 Raymond Damadian, of the Downstate Medical Center in Brooklyn, found that the relaxation time in normal mice cells was different from that in cancerous cells. In 1972 Damadian patented a procedure for identifying cancer in biological tissues. The experiments provided no 'spatial' information, indicating instead the existence of cancer cells but not their position in the organism under investigation. No one was able to tell from what place in the sample the NMR signal arose. But in 1974 Paul C. Lauterbur in the USA and Peter Mansfield in the UK described using magnetic field gradients to locate from which point the anomalous NMR signal emitted. It was a revolution, the beginning of what would become a mainstay of medical diagnostics, magnetic resonance imaging (MRI). In 1982, the Mansfield group reported the first real-time images of the heart of a living rat, opening the way for MRI's use in the diagnosis of congenital heart disease. Needless to say, in 2003 Lauterbur and Mansfield were jointly awarded the Nobel Prize for Medicine.

Today, magnetic resonance has replaced many forms of invasive diagnostic analysis, with great benefits for the patients. To achieve this, many decades of work by a large number of scientists have been necessary. Along with sudden jumps, there have been many minor improvements in the technique, each one contributing to make it available to all of us. But without Rabi's fundamental experiment, NMR would never have come to light.

This, in a nutshell, is basic science; asking those who are working in basic research for an immediate purpose may be useless, as well as inappropriate. But there is a problem. Not everybody is as good as Rabi, and for many who do basic research, possibly not endowed with special genius or innovativeness, there is a real danger that research will never lead to solid results. If genius and ability are totally lacking, one may end up grinding water in a mortar—wasting time and money, in short, being irrelevant. But this dark side must be tolerated in the name of greater benefits. However, there is a further, even bigger problem. The mechanisms that have led in the past to steady progress in scientific knowledge have partially deteriorated in recent years, and are beginning to show worrying crunches. Hopefully, it is too early to worry. But at least from the point of view of a long-standing, passionate researcher as I am, some of these twists are alarming signals that should not be neglected or underestimated.

2

The way we were: Doing science in the previous century

It is not easy to explain to somebody who has never experienced the environment of scientific research how it works. It is not easy because it is one thing to have a superficial knowledge of something acquired from reading books or watching movies, and a completely different thing to have direct and personal knowledge. It is not easy also because most of us would most likely offer a personal and restricted view of what science is, presumably distorted through the lens of our own experience. Moreover, there is not a single way of doing scientific research: there are many, and each 'scientific area', from engineering to medicine, from mathematics to biology, has its own habits, behaviour, traditions, and even rituals and obsessions. No doubt, the scientific life of a mathematician is extremely different from that of a cardiac surgeon. Therefore, I will not try to tackle an enterprise that looks, more than difficult, impossible. I will simply provide a point of view—mine, grown out of what is by now a rather long and rich experience covering almost four decades of activity in the world of research.

The Overproduction of Truth. Gianfranco Pacchioni.

DOI: 10.1093/oso/9780198799887.001.0001

Behind the Iron Curtain

It was 1981 when I moved to West Berlin, at the Institute of Physical Chemistry of the Freie Universität. After graduating, my activity was devoted to the theoretical study of new exotic objects, aggregates of a few atoms, called clusters, something that lays halfway between the world of single atoms and the more familiar world of solid substances such as bulk iron or copper. During my thesis I had developed a computer program. At that time I was still using punched cards, an archaic way of communicating with computers but operational, although with lots of drawbacks. A few months earlier I had published some results that aroused the interest of a Berlin professor who offered me a chance to work with him for a while. Prospects at Italian universities were not particularly attractive, so I accepted, not without some misgivings.

Once in Berlin, I used to spend at least an hour a day in the library. I liked browsing the various journals to read the latest scientific articles in my field. At that time there were then no more than four or five major journals worth scanning, namely, *Physical Review*, more oriented towards physics; *Journal of Chemical Physics*, the leading chemical physics journal; *Chemical Physics Letters*, same subject but accepting articles of limited length, called letters, with shortened publication times; *Journal of Physical Chemistry*; and a few others. At that time *Journal of Physical Chemistry* was being published every two weeks, and by the end of 1980 this amounted to no less than 4,000 pages of original articles for the year. It was already a rather large amount of information, but, obviously, many aspects and topics were of no interest to me and I did not need to deal with them. Only twenty years earlier, in 1960, the same journal was publishing one issue per month for a total of less than 2,000 pages of scientific articles for the whole year. For comparison, today that same journal has been divided into three parts, A, B, and C, each dedicated to a sub-sector of physical chemistry, plus a separate letters section. Each part is published weekly, or about 180 issues per year all together, for a total of roughly 60,000 pages. These bare numbers are sufficient to give a feeling of what is happening at present in the world of science.

But let's go back to Berlin. In those years, as I said, I enjoyed spending time in the library, and an hour a day was just enough to keep me up to date on the activities of the 10–15 groups worldwide who were tackling the same problems on which I was working. At that time, before the Internet era, all communication was via the mail. One would mail 3–4 copies of a new manuscript to a journal editor, asking for evaluation and possible publication. The editor would send the work to a reviewer, sometimes to two, the famous 'referees', a technical term of scientific research. The overseas correspondence (major journals were often American) was then taking two weeks, barring unfavourable circumstances. Then, referees would take their time to read the manuscript and to provide an evaluation, usually asking for a number of modifications if the paper was accepted for publication, or, if the quality was judged to be insufficient, rejecting it with due motivation.

Getting through the process could easily take several months. At that point modifications were made, again using only mechanical typewriters. In order to reduce the retyping of the manuscript to an indispensable minimum, changes and additions were made by typing up small pieces of text and splicing them into the main text with sellotape. Literally, a cut-and-paste operation, and the name remains today even when we do the same on a computer. Upon completion, the revised manuscript was sent back to the editor. At that point all one could do was wait (anxiously and patiently) for the response letter (again through the mail) with the final decision: acceptance or rejection. After this, in case of acceptance, galley proofs were sent out for correction months later because the manuscript had to be actually set in type for printing. Proofs were annotated manually, with a red pen, and sent back to the publisher. Finally, several months and sometimes even a full year after the first submission, if everything was all right the work would appear in a published issue of the journal, which was sent, again by mail, to libraries around the world.

A common practice at the time was that anyone who had read a scientific article of particular interest could send the author a special postcard, known as a reprint request, kindly asking for a copy of that article. Each university or research centre had its own reprint request cards, with an empty space to be filled in by hand with the title of the article, year, volume and page number, etc., and address of the author. In fact, after publishing a paper the author received a stipulated number of offprints, or copies of the paper, reprinted separately by the publisher that the author could distribute to colleagues. It was a tradition that originated in a time when photocopying machines did not exist, and the only way to get a copy of an article was to ask the author directly for a reprint. It was also an indirect way of measuring how much attention one's work was receiving: no or very few reprint requests, little interest; many reprint requests, great success! In short, it was the 'I like' of twentieth-century science.

On the western side of the Iron Curtain, I had been in Berlin for a few months when I received one, then two, then three reprint requests, all coming from the same person, a Helmut Haberland of the Academy of Sciences in Berlin-Adlershof, in the *infamous* German Democratic Republic, DDR for short. The Academy of Sciences of the DDR was only about ten miles away from my institute in West Berlin, in the beautiful area of Dahlem. But at that time, with the Wall in between, it was in another world, mysterious, inaccessible, and potentially hostile. After duly sending out the reprints after the first two requests, when I received a third one I decided to write a letter. Given his interest for my work and our relative geographic 'vicinity', I proposed to meet, offering to go to East Berlin for a visit, being aware that for them the reciprocal was simply impossible. Since 1961 West Berlin was enclosed by the (in)famous Wall, in practice a continuous, reinforced, off-limits border strip a few hundred meters wide, with barbed-wire entanglements, and control turrets manned by armed guards, and with a concrete wall about three meters high towering in the middle. It was the so-called Iron Curtain, as so named by Winston Churchill in a famous speech in 1946. We in West Berlin were the 'free' world, although captive and trapped in this enclave. But, thanks to my Italian passport, I was free to travel, to some extent, even though I could be subjected to a boring queue at the Friedrichstrasse crossing point, in the centre of East Berlin, for passport control and other formalities. Well aware of this possibility, I set forth my proposal to meet with my mysterious colleague in East Berlin and sent my letter in search of response. I waited weeks, and months, but unfortunately the answer never came.

One year later I went to Prague to give an oral communication to present the results of my research. I was very excited because it was my first participation in an international Congress. One evening, at the end of the session, I was approached by a cautious and circumspect person. He introduces himself as Dr Helmut Haberland, the individual who had sent the reprint requests. After the ritual compliments, I ask him if he ever received my letter and, if so, why he never answered. Poor Haberland must have thought he was in the presence of a very naive person, one of those Westerners who fill their mouths with claims of socialist society but have never had a chance of actually trying to dwell in such a system. He proceeded to explain to me that he cannot send letters in sealed envelopes to a stranger, particularly in West Berlin; that such messages would be opened and examined, most likely with adverse consequences for senders. I could hardly digest such news, nor could I imagine that things were really at that point. Then I proposed to arrange a meeting as soon as we both returned, suggesting to meet in East Berlin, to have a coffee or a beer together, and to talk about our work. 'This guy does not seem to understand' must have been Haberland's impression. Impossible, he explains patiently, as if talking to a small child: if we meet 'privately', the fact will immediately be reported to Stasi, the powerful and dreaded secret police of the DDR, for sure causing even worse trouble. At that moment I began to realize how much of the facts of common life behind the 'Iron Curtain' came through to us Westerners only in a blurred and screened manner.

Haberland, however, was genuinely interested in establishing a contact with me and promised that he would seriously endeavour to invite me, this time in an official way and with all blessings, for a visit to take place the following year at the Theoretical Chemistry Conference of the DDR. And, lo and behold, a few months later I was permitted to go for a couple of weeks to the DDR in Heiligendamm, where the Congress took place, to give a communication and to visit with some research groups in East Berlin and in Jena. It was a memorable visit, in every sense. The atmosphere out there, on the other side of Wall, was quite unique and particularly oppressive.

The 'scientific' visits were always preceded by formal conversations with political personalities (in decreasing order of importance): the Director of the Academy of Sciences, the Director of the Department of Physics, the Director of the Theoretical Chemistry Laboratory, and so on, until eventually I was able to speak with the people who were actually doing research. Among them were a young couple who for various reasons seemed to be very interested in my work. They worked in the same institute as Haberland and were also working on theoretical models of molecular systems. The gentleman's name was Joachim Sauer, and the lady's name was Angela Merkel, who were to become known as the future Chancellor of the Federal Republic of Germany and her faithful husband. But at that moment the thought that

East and West Germany would one day be reunited was simply unimaginable, a pure fairy tale. Even more out of this world was the idea that in front of me stood a colleague who would one day become the most powerful woman in the world.

Be that as it may, I remember one occasion when I spent an hour or so with her sitting in a coffee house in East Berlin, talking about how to cook artichokes rather than about major social or political issues—something that in 1982 nobody would have dreamed of touching upon in a friendly conversation, in a public place such as where we were! With Sauer and Haberland, we found much common ground for striking up a collaboration. Beside the scientific aspects, for these friends it was also extremely important to stay in touch with someone from a Western country, an absolutely unique chance for them. Apart from the obvious scientific reasons, that contact has had for me a special flavour of the mysterious, secret, and forbidden aspects associated with it. It was also a rare opportunity to become acquainted with a world politically, socially, and physically separated from ours.

So we began a regular scientific collaboration. People in East Berlin had developed some promising theoretical models, but had at their disposal only obsolete and primitive computers to test them; in contrast, I had access to the powerful research tools of the Western world. We exchanged results and impressions in long typewritten letters, which were travelling at an awfully slow pace through the thick and impervious border measures. That correspondence had in the meantime been authorized, although it was one in which I had learned to strictly talk about nothing else than convergence criteria, potential energy surfaces, chemical bonds, and other similar technicalities. From time to time, say every 4–5 months, I managed to make a one-day trip from 'here' to 'there', in order to meet in person. I had to board the underground at a West Berlin station and ride it into the dark gloom of East Berlin, crossing through ghostly stations manned by guards toting heavy machine guns. Those stations were abandoned and were unused since the famous date of 13 August 1961, the day construction of the Wall started. I would leave the train at Friedrichstrasse where there was a checkpoint, and I had to climb a badly lit stairway in a sort of no-man's land where I was no longer in West Berlin but had not yet entered East Berlin. Since I had an official visa, I was sent to a separate counter where there was never any queue, and I was subjected to some ritual checks (sometimes routine, sometimes less so, but that is another story). On the occasion of my first visit to East Berlin, once I had passed the checks there was a guy waiting for me with whom I had no particular business: Dr Gay was his name, and he was supposed to escort me on the S-Bahn trip to the Academy of Sciences. Later, in a private location and far from indiscreet ears, I learned that Dr Gay was a *reisekader*, that is, a person with a permanent permit to go abroad. As everybody was well aware of, this permit was issued only to people who agreed, often upon some sort of blackmail, to collaborate with Stasi and provide regular reports on contacts, people, relatives, and acquaintances. Dr Gay was there to check that I was not a spy and with him I could only talk about soccer and *spaghetti alla carbonara*. After three or four visits somebody realized that as a spy I was simply a disaster, and from that moment on I found Haberland waiting for me, which was much better because at least we had something to talk about during the boring trip to Adlershof.

In the long run we published some papers together, which turned out to be truly pioneering studies in our field at the time. But between the first drafts of the papers and their publication, years passed, with letters sent, long periods of silence, reply letters received (telephone contact was not allowed), manuscripts going back and forth with handwritten annotations, and so on. It was a nineteenth-century way of doing science, I liberally admit, and it might have seemed a little anachronistic even at that time. That collaboration went on for a few years, and when I returned to Italy, I made a great effort to invite one of my Berlin friends to Milan. I sent out personal letters of invitation, others were sent by the director of my department to the director of their institute, all in vain. However, one day in June 1989, the phone rang on my desk. At the other end of the wire was Haberland himself, and this was already an absolute novelty. They had been given permission to go abroad, he told me, he could come to Milan; he was truly excited. He had waited for that moment ever since August 1961. I was excited too. In the following months we went about arranging the details of the visit: when, how, and all that. And then 9 November 1989 arrived, the day and event no one could have anticipated just a few weeks before: the crumbling of the Berlin Wall! Haberland came to visit me at the beginning of 1990, but, in fact, by then the DDR no longer existed.

California, The Golden State

If collaborating with colleagues in East Berlin made me aware of a world of technological backwardness, then going to work for a few months at an IBM research centre in Almaden, California, in 1987 was a journey into the future. While I was in Berlin I had the opportunity to meet an IBM scientist who was considered an icon in our field, providing background work for his seminal papers, papers that I read and studied with great attention. He was one of those scientists whom young people (as I was then) approach to receive valuable advice or even just approval for the work they have been doing, perhaps in the remote hope of establishing a contact that one day can turn into a factual collaboration. This is what happened with the American scholar whom I was able to contact thanks to a fellowship from IBM's Italian branch, and whose name is Paul Bagus. The jump from the Academy of Sciences of the DDR could hardly have been more radical and shocking. As much as East Berlin's environment was crummy, muddy, dusty, and restricted in terms of infrastructure and instrumentation, Almaden was one of the most advanced temples of contemporary science. In those years, every researcher's dream was to work in one of America's major industrial research centres: the IBM ones (there were only two in the United States: one in Almaden and one near New York) and Bell Laboratories.

At that time I had only heard about these mythic places, but once there I soon realized that their reputation had a solid foundation. The environment was sparkling, the labs were almost science fiction, the computers were super-powerful, the library was immense, the cafeteria was competing with fashionable coffee shops in downtown London or Paris. In the corridors, in the cafeteria, in the library, it was not uncommon to cross paths with one or another Nobel Prize winner, or with some great scientist whose contributions one had been studying in so much detail. People of all races and nationalities were coming and going, having in common only a deep interest and involvement in basic research and pure science.

The work at IBM was making intense progress and giving positive results. I was interacting a lot with Paul; every evening he would come to my office to discuss the day's progress, and we would have long discussions about the results obtained, or problems encountered, planning new calculations, formulating reasonable interpretations and hypotheses. We theoreticians often met with experimentalist colleagues, always intense, informed, sometimes even animated. Everything was aimed at understanding a new phenomenon, interpreting an experiment, testing an original hypothesis. Every now and then we would discuss the possibility of publishing our results (I was definitely the most interested, as I was anxiously looking forward to reading my name in print with the affiliation 'IBM, Almaden, California').

But there was no particular pressure nor hurry, and in fact the first paper I published with Paul appeared in 1989, two years after my first visit. I worked with passion and dedication, with the purpose of doing something new and of lasting value. I was exploring unknown routes, without forgetting the importance of transforming the work into a published article, which is invariably the tangible evidence of a researcher's scientific progress, although without overwhelming pressure. In this respect, I must say that the attitude at the Freie Universität, at the Academy of Sciences of the DDR, and at the IBM centre in California was definitely the same.

Nor, for that matter, was the time required to publish a scientific paper very different. Of course, major technological tools were available at IBM, as well as specialized staff assisting researchers to perform their work, such as expert lab technicians, or people specialized in producing manuscript drawings and pictures, or highly efficient secretaries. One still had to go through the standard procedure of posting one's manuscript, after several drafts and different versions, to the most appropriate journal, and referees reports, editors' responses, and so on, had to be exchanged, always by mail, with authors. Collaborations with other groups, though facilitated via the telephone and fax and by opportunities to meet personally even if that involved expensive flights, were still mainly based on paper copies, with formal epistle style. All of that had a very special consequence: although carried out intensely, and with dedication and strong involvement, research progressed with allowance for idle times that left room for reflection, doubts, reasoning, and discussion. The slowness of mail correspondence provided, in fact, spare time to think, to elaborate, somehow compelling researchers to reappreciate longer the conclusions of a given work, resulting in more extensive, comprehensive, and accurate written accounts. We will see in what follows that things have changed dramatically in this respect.

'Deutschland über alles'

In 1993 I made another excursion abroad, this time to Munich, at the Technische Universität. In Milan I was always short of funds and of computing resources for my research. Collaboration with groups in other countries was therefore a matter of survival. The business worked more or less as follows: I provided some ideas that I could not exploit alone due to the shortage of means, and someone else provided the infrastructure—computing in this case. Actually, there was more to this arrangement. Each new experience brings along new knowledge, different working methods, personal relationships, and also alternative ways of performing research. In Berlin I worked directly with my boss, without much intermediation, except for interactions with other members of the group. The link at IBM was in practice only with Paul Bagus, in what may be called a very one-to-one relationship of a mentor with his pupil, but it worked out greatly. When I arrived in Munich, I found the kind of organization that today forms the basic gear of modern research: a professor, heading a large lab with some experienced coworkers, the so-called post-docs or post-graduates. These are typically people aged between 30 and 35 who already have a good deal of past experience, but do not yet have a consolidated position in the academic world. Each post-doc was supervising a few doctoral students who in turn could be responsible for a master student. Last but not least, there was an indispensable and energetic secretary.

In other words, it was a pyramid of authority, with the pharaoh at the summit who, as in ancient Egyptian dynasties, had power of life and death (in an academic sense, of course) over his 'workers'. It was my first contact with a top-down, hierarchical structure and organization, with all its associated pros and cons: jealousy but also solidarity between group members, internal competition but also allocation of tasks, and so on. The head was Notker

Rösch, a not easy character but an expert and brilliant scientist with the typical attitude of a Prussian general toward his unruly troopers. I was a guest visitor, and as such less obliged to toe the line, but not entirely relieved from 'keeping rank'. On Friday afternoons there was a meeting intended for joint discussion of individual problems and to find their proper solution, to report on the latest results of other 'competing' groups, and other such matters. In practice, the meetings were prone to turning into a monologue of the boss, whom, of course, no one dared interrupt, and lasted until the exhaustion of the attendees (but not of the speaker). For the first time I was working with a really large group, 20–5 people, sometimes even more, and I realized how challenging the assignment of a lively theme and of an original scientific problem can be for each member to deal with.

In some research fields students are 'beasts of burden', indefatigable workers dedicated to synthesizing an unknown substance, measuring an exhausting series of samples, building a small piece of apparatus, or developing a new subroutine of computer code. In such cases the idea is unique, and manpower contributes lots of work and patience to its accomplishment. In other fields, research is rather individualistic and everyone has a different problem to solve, more or less broad and complex. In Munich, I became aware of how so many collaborators can be engaged on topics perhaps not entirely original, but very 'systematic', meaning that the same problem is tackled in different ways, in order to record differences, improvements, and changes.

An instructive story

In those years we were engaged on a topic whose story is worth being told.[4] By 1990 I had begun to deal with the interaction of a small molecule like carbon monoxide, in chemical formula CO, and a solid material known as magnesium oxide, or MgO. About ten years earlier, some experiments investigating how MgO's surface interacts and absorbs the small CO molecules appeared in the literature. The interest was dictated by the fact that carbon monoxide, a poisonous molecule, reacts with many substances and transforms into other chemical species if activated by particular systems known as catalysts. Magnesium oxide is an excellent model system of these catalysts, as such prompting interest in the problem. The experimentalists determined the strength by which the CO molecule binds to the MgO particle surface, and provided a value of 0.15 eV, where eV stands for electron volts, a measure of energy, just like the meter is a measure of length.

When I had collaborated with Paul Bagus at IBM, I decided to look into the problem and proposed an electrostatic model for explaining the interaction. Once in Munich in the Rösch group, we decided to reconsider the problem using a different and rather new theory, but surprisingly we found much larger values for the interaction energy between CO and MgO, about 0.5 eV, hardly reconcilable with the experimental data. It was a clear sign that the method was not accurate enough. In fact, a couple of years later, with some corrections that took into account effects not considered in the first model, we obtained a value of 0.1 eV, rather close to the original experimental data. It should be mentioned that at that time I blindly and uncritically believed the data that came from experiment: these are measured, I thought, therefore must be undisputable. Actually, things are not always so and, as we will see, experimental data must be interpreted and understood, and they may change over time thanks to improved techniques.

Still, after our last work, the question was not entirely solved as it was thought that the original experimental data, those that had produced the famous value 0.15 eV, were partially affected by poor control over the magnesium oxide particles on which the measurements had been made. Ideally, these should have been perfect cubic crystals, but we suspected that they could have contained various kinds of defects that inevitably made the experimental measurements less precise. A leap forward occurred in 1993 when a world-leading group at Texas A&M University developed a new method for preparing magnesium oxide surfaces. It was a technique that made it possible to 'spread' a thin layer of magnesium oxide on a support (a metallic surface), creating a kind of ideal, defect-free surface. With the first application of this new system, the experimental group in Texas managed to absorb carbon monoxide and to measure the strength of the bond. The value they obtained, 0.35 eV, was more than twice the previous experimental value, and also much larger than the most accurate value we obtained from the theoretical calculations.

As usual in these cases, one first needs to understand what is not working in the theoretical model and how it can be improved. The problem had become 'hot' and various groups around the world began to calculate the strength of the bonding interaction between carbon monoxide and magnesium oxide, using more sophisticated methods, more complex and elaborated calculations, and more extensive models. The most advanced techniques available were applied, while new ones were developed, with great investments of time and manpower. But with only one, disappointing result: as the theory was refined and improved, the value of that critical interaction energy decreased, instead of increasing and reaching, as expected, the much-coveted target of 0.35 eV, the latest experimental measure. Indeed, the most accurate theoretical data predicted values below 0.10 eV, quite far from experiment. At that point some frustration began to circulate. Several theoretical groups were unable to reproduce some experimental data of a well-defined, apparently simple system. They could not even approach the reference value, and, even worse, they could not understand why. I will let you imagine how many fanciful hypotheses were proposed, without success, in the attempt to make ends meet.

At that moment a colleague of mine from the University of Stockholm, Lars Pettersson, questioned the experimental measurement made by the Texas A&M group. It should be recalled that if one adsorbs a molecule onto a flat surface, smooth as a table shelf, the interaction will have some given value, but if the surface for some reason is rough, with atomic steps or irregularities, then the interaction will be different, usually stronger. Obviously, everyone was convinced that the samples prepared by the Texas A&M team were as smooth as a mirror, since this is what they claimed to get with their new technique. Lars calculated how strongly the CO molecule would bind to the smooth and rough surfaces of magnesium oxide, and concluded that the Texas experiments were made under conditions that allowed the CO molecules to link only to the rough surface, where the interaction is stronger, and not to the flat, regular surface that theoreticians had been considering in their models. In other words, the values did not match because the experimentalists were measuring one condition, whereas the theoreticians' calculations assumed a different one. It was an uncomfortable truth, difficult for the experimental community to accept. It implied that the produced magnesium oxide samples were by no means as 'perfect' as had been assumed. This result, if confirmed, would make the new preparation method less appealing or at least establish a limit to its use.

For people active in research, this kind of problem is an everyday one. Nobody 'sees' directly what is prepared or synthesized. Evidence is almost always indirect, in the sense that to characterize a new sample, whether biological or inorganic, it is necessary to resort to external responses and signals whose interpretation is not necessarily straightforward. Until the end of the past century nobody had ever 'seen' atoms, but nobody questioned their existence. Many things are known to exist or are done in a certain way without direct visual reconnaissance or proof. That was also the case for the thin magnesium oxide layers prepared at Texas A&M. However, there had been no room for convergence: theory had pushed its potential to its limits without reaching the experimental value and actually called the experiment into question; experiment had produced a result, leaving theoreticians with the problem of its reproduction and interpretation. It was a real impasse, to the point that for a couple of years no one made any progress.

At this stage an event occurred, still vivid in my memory, that in a way introduces us to the problem of the distortions affecting contemporary science. It was summer 1997 and I was visiting a theoretical chemist friend at the University of Barcelona, Francesc Illas. We were at the beginning of the Internet era, and accessing journals in electronic format by downloading them online was a rather novel opportunity. I was intent on browsing the latest issues of some journals when my attention was drawn to a paper that had appeared not long before in *Chemical Physics Letters*. An author of Chinese nationality, unknown to me, was reporting a study on the topic of CO–MgO interaction using theoretical models. To my great surprise, the paper claimed to have solved the problem, with new calculations that reportedly had achieved a value very close to the much-publicized experimental value of 0.35 eV: their result was 0.4 eV! Obviously, I rushed to read the article, being very keen to discover what magical advance had escaped all of us for all those years. It looked like, I had to admit, that we had been outdone by an obscure Chinese outsider!

As I read further, however, amazement and disbelief began to grow: the Chinese scientist had done nothing else but use the old method introduced by myself and Rösch in Munich a few years earlier, the one that provided a binding energy of 0.5 eV.

But we knew that this was an overestimate, and we also knew very well why, as reported and explained in several subsequent studies. Even more surprising was the fact that in the new paper the author ignored some of the papers that had appeared in the preceding years, not even citing them, whereas others he was apparently aware of, but had deliberately decided to ignore their contents in drawing his conclusions. Using a method others had already shown to be inadequate, the Chinese researcher had found good agreement with experimental data and therefore concluded, repeatedly, that his approach was the correct one, that all the improvements introduced in those years were useless, and, most important, that the magnesium oxide samples produced at Texas A&M were smooth and clean as a glass plate. The consequences of this work were, apparently, really devastating. On the one hand, the work dismissed years of effort made by various groups to improve the theory; on the other hand, it provided theoretical support to prove that the preparation technique of the Texas scientists produced perfect surfaces: exactly the opposite of what many of us were claiming.

In modern science when this kind of controversy arises, it may not be easy to sort out. Parties and lobbies start to form, some in favour of one side while some stick with the other. Even while the corners of the dilemma remain the same, long and inflamed discussions and diatribes develop and can last for years. More than that, what was at stake was of far greater importance than the relatively simple problem from which everything had started. It was no longer just a matter of obtaining a precise value for the interaction energy between CO and MgO, but also of deciding which experimental method should be used to prepare these systems, and which theoretical method is best suited to describe them. The case had ended in a blind alley, and to clear out of there a fresh reconsideration and a substantial step forward were definitely needed.

End of the dispute

This is what happened, thanks to Hajo Freund's experimental group at the Fritz Haber Institut in Berlin. Freund headed a group of more than 50 people and could afford to devote someone of them to attempt a delicate experiment with few chances of success. Until then, the interaction between CO and MgO had been measured either on tiny crystallites, as done in the 1980s, or on very thin layers, as done by the American group. What was needed was a similar measure but on a so-called 'single crystal', that is, a single block of material with perfect crystalline structure several millimeters in size. Single crystals are, for example, those used for diamond jewels and other precious gemstones. As with diamonds, individual crystal faces can be cut off with particular sides that are very smooth and regular. For various reasons, however, experimental measurements on these crystals are particularly tedious and difficult. It took more than a year of work before the result that finally clarified the matter could be achieved: the interaction energy measured on a smooth magnesium oxide surface was 0.14 eV, close to the more accurate theoretical values, and far from the value of 0.35 eV reported years before by the Texas A&M team. In addition, with other experiments it became possible to demonstrate that what the American group had measured were indeed CO molecules attached to steps and defects, which evidently were present on the samples, despite the initial claims. Theory was safe, and indeed had contributed decisively to bring to light a problem inherent to the measures taken in Texas. The circle was closed, the problem solved. But it took almost ten years of joint efforts, discussions, hypotheses, and new experimental design.

If you were to ask me what is the benefit of knowing precisely the value of the interaction energy of carbon monoxide with magnesium oxide's surface, the answer would be simple and lapidarian: it is rather useless, it's just bare data. But I hope I have given an idea of how the work performed to provide an answer to a problem of apparently low relevance can stimulate efforts and interests, and in the end contribute to advance the field, both experimentally and theoretically. But one more consideration must be made. Without the Berlin experiment, the solution to the problem would have taken a long time. Sooner or later, the truth would have emerged because, and this is the strength of science, poorly interpreted or wrong results do not withstand robust verification and eventually are identified as such. But it is also true that the work of the Chinese researcher who had deliberately ignored much of what had been done previously, leading to opposing conclusions, could have done a lot of damage. Particularly surprising is the fact that such a paper was finally published in a journal of good reputation, passing the verifications carried out in the review phase. This was a clear case, albeit unfortunately more and more frequent, of the failure of filtering at the peer review process level, which is the basis for validating all studies that claim to be scientific. It was also a typical case of carelessness and of the drive to publish scientific work at any cost, even if reporting results that would later prove to be not only wrong, but also potentially harmful.

After reading the paper in my office in Barcelona, once I recovered from surprise and outrage, I immediately wrote a rebuttal against the recently published data and, thanks to the Internet, I shared it in real time with those who had been contributing to the problem. A letter was published later in the same journal where we made clear the severe limitations of the Chinese work and tried to restore a scientific truth. For the first time I was confronted with such an obvious case of twisted science: pressure to publish new but incorrect results; unfairness in citing improperly, or not citing, previous work; clear failure of the peer review process; the appearance in the literature of a paper that was not only unnecessary, but could actually have been detrimental. All this in one shot. It was a harbinger of what, unfortunately, was to become a recurrent phenomenon.

3

Publish or perish

As a young man I got involved in athletics. I remember long afternoons at the stadium, and team championships based on scores awarded for each athlete's performance: 100 meters in 11 seconds, 960 points; 5000 meters in 14 minutes, 1020 points; and so on. Finally, the points scored by all the athletes of a given team were summed up to generate team rankings. That was all. The world of sports measures performance in perhaps the simplest and most direct way. To establish the rank of a high jumper, you look at his numerical results, that is, his personal record, how many times he has jumped a certain height at the first attempt, and so on. All gathered information are objective, measurable, and unquestionable data. If we wanted to build a ranking of the best soccer players, things get a bit more complicated: we could count how many matches one has played in the top league, or his appearances on the national team. For a centre forward, we could also count the number of goals scored, but for the backs this would not be a reliable indicator. It is even more difficult for a goalkeeper; however, we could record how many shots to his goal he has stopped, but then we would need to say how many were opponents' kicks and how many were just a backward pass from a teammate.

And how about evaluating a scientist's curriculum vitae? Well, here the matter is far from simple. At the basis of a scientist's success lays a fundamental fact, the amount of scientific work he or she has been able to publish. Obviously, this parameter too

The Overproduction of Truth. Gianfranco Pacchioni.

DOI: 10.1093/oso/9780198799887.001.0001

is inaccurate, because as with a goalie, where not all catches are equally difficult, so for a researcher not all studies are equally relevant. We will see in what follows that, in fact, there are many other indicators, and how they lend themselves to manipulations and distortions if used in an improper way. But at the bottom of it all, there is and still remains scientific publication. Without it, other indicators lose effect, like a plant without roots.

In the past, the publication of a scientific work, or its presentation before a scientific society, was the end of a long and tiring journey. It could take years before one came to write a paper describing experiments, observations, and conclusions, perhaps within a new interpretative model. Darwin published his epochal book *On the Origin of the Species* on 24 November 1859, but he had started working on his theory already during the famous HMS *Beagle* tour in the 1830s. Darwin continued to accumulate data and to carefully polish his theory, aware that in order to question creationism, his theory had to be 'absolutely watertight'. In January 1842 he sent a description of his theory to his colleague Charles Lyell. In the spring of 1856, 14 years later, Lyell showed Darwin an introductory paper about the origin of species by Alfred Wallace, another naturalist working on the same subject. Lyell tried to convince Darwin to publish his work, not to be preceded by Wallace. In May 1856, Darwin decided to set down to work in earnest, in order to present a complete and coherent text of his theory, which he did three years later. From the first observations to the publication of what remains a milestone in the history of modern science, almost thirty years had passed.

How much do we publish?

There was a very small number of protagonists in the world of science in Darwin's time, and the probability that someone would be working on the same themes and could thus precede someone's presentation of similar results was rather low (but not zero, as we have seen). With the great revolutions of the past century in all scientific fields, from physics to biology, from chemistry to medicine, the number of scientists and scientific publications began to grow, initially at a low rate, and then gradually at an ever-increasing rate. At the time I started my 'career', I remember that older and well-respected professors had usually produced in their scientific life just a few dozen publications in specialized journals, often of only national circulation. Publishing one or two papers per year was considered an intense production. Then, slowly, over the years, there has been a steady increase in frequency, a growing scientific production, and the number of published papers per author has increased, to become recently a true avalanche. This has resulted in a strong and growing pressure to publish results in specialized journals, giving rise to the well-known effect in the science community (and not only there) of *publish or perish*. One could hardly find a more effective expression. The number of published scientific writings (e.g., articles, books) has quickly become the main (though not the only) factor for assessing a researcher's performance. As already said, not all papers have the same importance and impact, and not all 'discoveries' are at the same level of innovation or relevance. That is why other 'measuring' tools have been developed, to which we will return to later. These tools have played an important role in setting the direction that modern science has taken.

Derek J. de Solla Price (1922–83), a British physicist and historian of English science, was the first to address the issue of the mass of scientific publications and its evolution over time. In 1963 de Solla Price published the first quantitative analysis of the way scientific knowledge has evolved from 1650 to 1960.[5] In 1960 about two million scientific papers were available to the world community, having been produced over three centuries. It may seem an impressive number, but we must consider that de Solla Price's study included all scientific disciplines, which encompasses a rather vast period of time. The interesting aspect, though, is that most of this mass of results was produced between 1910 and 1960, while the number of papers produced in the previous two and a half centuries was on the order of a few thousand or tens of thousands at most, as shown in Figure 2.

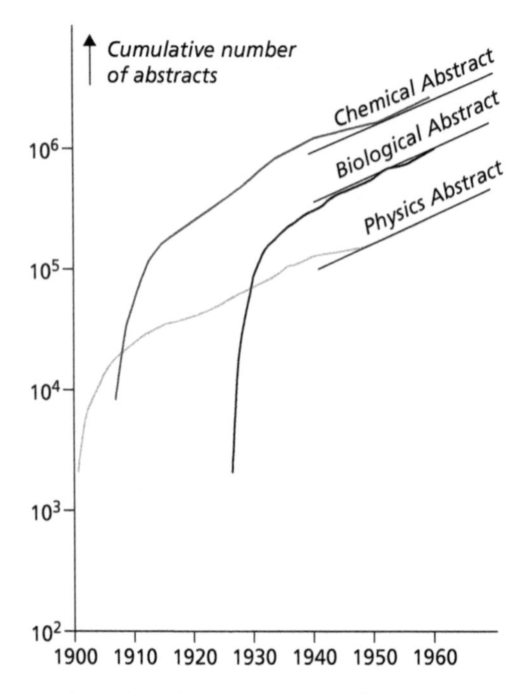

Figure 2 Total number of abstracts published in various scientific fields. The data are collected from the moment that collections of abstracts of scientific works appeared in the literature.

Source: de Solla Price, D. J. (1963), *Little Science, Big Science*, New York: Columbia University Press.

The 300 years from 1650 to 1960 saw radical advances in the world of science, and some real revolutions: from thermodynamics to electromagnetism; from evolution theory to the quantum world and to the theory of relativity; from the discoveries of molecular biology, including the DNA structure, to the foundations of genetics; and so on. In short, along with that multitude of scientific articles, there stands a consistent and tangible array of fundamental advances in science and technology that have profoundly changed the world. Although we cannot say that all these works have played a decisive role in modern scientific development, it is likely that most of them made steps, albeit small ones, towards global knowledge. Common belief says that things are not necessarily so when we consider present times. For example, today the number of scientific articles published each year and passing through a more or less strict peer-review process stands between 2 and 2.5 million.[6,7] This means that in one year the amount of scientific information produced exceeds, at least numerically, everything that had been generated over the three hundred years from 1650 to 1960. The consequence is that the Web of Science database includes about 90 million records while Google Scholar is estimated to index between 100 and 160 million scientific documents.[8,9]

Despite the fact that scientific and technological development is increasing very rapidly today, it is hard to believe that each of the two million papers published this year will bring definite and measurable advances in the scientific world. The pressure to publish, as we shall see, leads to undesirable phenomena, with far from positive consequences for scientific development.

There is, first, a practical result from the relentless increase in scientific knowledge that is both obvious and disconcerting: it is literally impossible to read and follow everything that is produced. In a bold preface with a prophetic name, *Exponentials,* published in 1968, chemist Paul B. Weisz went as far as stating a mathematical consideration whose paradoxical conclusion we can only fully share.[10] Weisz observed that the rate of production of scientific work in the 1960s was already growing at an exponential rate. Currently, the total number of existing articles doubles every 8–9 years, according to the latest estimates.[11] Countering this growing mass of information, Weisz pointed out that the time it takes for a researcher to read a given number of articles remains essentially constant. Thus, Weisz concluded, the percentage of information that a scientist can read and 'digest' is a quantity that decreases exponentially over time, being the result of the ratio between a constant (the number of scientific papers we can read) and an exponentially growing function (the number of papers produced over the years). As much as it is disturbing, this is a very proper conclusion, which leads to a very serious question: do we really know all the findings, results, and data produced even in the narrow and specific field we are dealing with? The answer is simply: 'no'!

Mega-collaborations

One reason, not the only one, for which the number of scientific papers increases steadily is that the opportunity of communicating in real time via the Internet has greatly enhanced the chances for collaborations. Collaboration is a staple of science and it is mirrored by the number of authors and institutions that contribute to a given paper. In 1988 only 8 per cent of all articles were based on international collaborations; twenty years later, in 2009, this fraction had already risen to 23 per cent. And in heavily technological countries such as the United States and those in Europe, the number of papers with international co-authors ranges between 27 and 42 per cent, according to a study of the National Science Foundation.[12] The UK's Royal Society estimates that today 35 per cent of all scientific papers are based on international collaboration.[13] Inevitably, the average number of co-authors on each paper has grown over time. In the United States, it has increased from 3.2 authors per publication in 1990 to 5.6 in 2010 (see endnote 12). Things can be very different in different research areas. The highest numbers of co-authors are found in the fields of physics and astrophysics, while the tendency for individual work is still very strong in mathematics and social sciences.

There is, however, an aspect of this phenomenon that is beginning to take on a disquieting size. I refer to hyper-collaborations, that is, works with more than 50 authors, and sometimes even more than 1,000. The term *hyper-authorship* was coined for such cases. In 1981, more or less when I started my scientific career, the highest number of authors on one paper was 118.[14] Since then, this number has grown to reach quite impressive peaks, as we shall see. The phenomenon is largely connected with the big international experiments of high-energy physics, but not exclusively. A strong push in this direction was given by the Large Hadron Collider (LHC), the CERN particle accelerator. A very large number of researchers, engineers, and technicians are involved in these mega-experiments. When CERN scientists announced they had finally observed the elusive Higgs boson, they published the result in *Physics Letters B*. The experiment involved more than 5,000 people altogether, and each of the two articles that disclose the findings is about 30 pages long, of which 20 are taken up by the list of involved names and institutions.[15,16] Both works established a record of about 3,000 authors each. Many questioned provocatively whether anyone would ever be able to break this record.[17] The answer came very soon. In 2015, a paper published in the prestige journal *Physical Review Letters* included 24 author pages in alphabetical order, for a total of 5,154 names and 344 institutions.[18] Would this article be fit for citing in a footnote?

Publishing in collaboration has the obvious consequence that productivity increases. The ATLAS experiment, which was the focus of the *Physics Letters B* articles mentioned above, is a striking example. All of the published papers related to this experiment report the authors in alphabetical order, so that the first author is a person named G. Aad. Between 2012 and 2016, that person authored more than 500 papers, namely, 123 papers in 2016 alone, nearly one every three days, Saturdays, Sundays, and other civil or religious holidays included. In these circumstances, even just reading all the articles in which your name appears becomes a heavy commitment. Anyway, there is no need for large teams of co-authors to achieve an impressive productivity in terms of published scientific articles. Recently, I became aware of the case of a pharmacologist working at an Italian university whose individual production is truly astonishing. Although with far fewer collaborators than high-energy physics experiments, this researcher, listing more than 1,350 publications at the time of writing, published 136 papers in 2015, and 146 in 2016, at a rate of three per week, if we exclude Easter and Christmas. It is a typical case where it is proper to wonder whether the author is aware of all that is written in his articles.

Start from scratch

Some years ago I had an opportunity to deal with a topic that was absolutely new to me. A physicist colleague prompted me to study theoretically the nature of some defects in a very common and relevant material, silicon dioxide. In fact, there were a great deal of data still to be understood and interpreted, and the computational models that I used could have provided useful answers. I tried to find out what was previously known in the field, a basic premise to begin my research. Silicon dioxide, SiO_2 in chemical formula, is of crucial importance in two key modern technologies: it is the main component of optical fibres, the ultra-thin glass wires that we use to exchange digital information; thin layers of SiO_2 were also the basis for the development of transistors and integrated circuits, which, in turn, enabled the development of modern computers. This means that without SiO_2 there would have been no electronic era, no Internet and all that comes from it. In addition, the application of fibre optics and transistors requires very high control over the defects that can be generated in this material. I soon became aware of how relevant this was, because digging into the literature of previous years and decades, rather than providing me with a clear and consistent picture of what was known, forced me to engage in a seemingly never-ending literature search.

I had to go to the library (the Internet was at its very beginning and there was no electronic access to journals), find an article, read it, identify 5–6 interesting bibliographic references, climb again and again on high ladders to retrieve dusty and heavily bound books, photocopy the articles in question, read them, mark other important references, and so on, in a complex, apparently endless game of cross references. After gathering some information I began to calculate the properties of some of these defects and to simulate their absorption and emission spectra.

As the work went hand in hand with reading vast amounts of collected articles, I began to understand the problem better and better, and to realize how to handle it. It took me more than a year before I could classify, read, and digest over one hundred scientific papers on the subject, having browsed and consulted more or less twice as many studies. At that point the bibliographic references bounced back and forth from one known paper to another, and finally I felt that I had collected the most significant facts about my problem. At that point I realized that these works came from very different time periods, from the 1950s onwards. Over that stretch of time, measuring techniques had been refined, new ones had been introduced, and the same problems had been revised in light of more accurate experiments. Thus, I decided to carry out an exercise that turned out to be very useful: I started reviewing all these works (now much faster, having read and annotated them) but in a strictly chronological order.

This made me understand better many more of the things that I had read in a random way in the previous months, until I got a perception of how the field had developed over the past four decades. At that moment I was ready to rigorously compare my new data with what was known in the literature. It had been a very massive effort (physically, too: the 'climbs' at the library were countless), but this allowed me, in a year's time, to become an 'expert' in that field. Today, twenty years after that stage of my career, if I had to work on that topic again, I would start from where I left off, and digest a couple of decades of literature that in recent years I have only been able to follow in passing. This, however, takes time, and time is at a premium today. As noted by Weisz, for all the efforts that one can put in place the ability to learn by studying is infinitely inferior to the evolution of knowledge. How many hidden things exist that I have never found out, and that could help me in my tasks? An unanswered question which leaves me with a sense of inadequacy, frustration, and a bit of bewilderment.

Law of survival

There is an unfortunate consequence of the overproduction of scientific work, in particular affecting the next generations of beginner scientists. In recent years, due to the great communication speed and to the speed at which things become obsolete, a trend that is seriously undermining some of the methodological foundations that have been the basis of scientific research until some years ago has developed. Faced with the obvious impossibility of reading, understanding, and digesting what has been done in the past, even the most immediate past, many young researchers, pressed by the tough and inflexible law of 'publish or perish', decide that it is simply useless to unearth and to appreciate what is already known. Instead, it is better and faster to 'rediscover it', *ex novo*. After all, one can quite understand why: studying a topic in depth would take months dedicated to searching and reading papers buried and scattered in the sea of scientific literature. Worse, this would severely slow down the original part of the research work, reducing the publication flow, and in the end, inevitably leading to 'succumbing', in the sense that a career, especially that of a young person, might be severely hampered. More and more often we witness the publication of studies where 'novel' results have actually been well known for years; the experiments' descriptions do not go beyond the paths that had been travelled long before; and long-established methods and procedures are proposed as original and innovative.[19] The projection to the future is so pronounced that a paper published five years ago is already considered old, a ten-year-old paper is considered obsolete, and if it is over 20 years old it is tagged for the dustbin or for the wax museum. However, this is quite unfair: old studies often contain a lot of insight, and new studies simply seem to reinvent the wheel—maybe with a large drumroll and trumpet blast, but still reinventing the wheel. This attitude goes along with one of the evils the scientific community has been struggling with for a long time: the credit for priority in a discovery, the merit of a new idea, the originality of a proposal.

What counts in the world of science is who comes first to a given idea or conclusion, which translates into the rigorous practice of citing previous work from which each researcher starts to aim at new targets. But if one omits to mention a published work in which similar ideas are reported, a superficial or uninformed reader may well assume that the discovery is new, original, and unprecedented. In reality, we are facing ignorance about what has been done in the past (an excusable misdemeanor, but nevertheless a misdemeanor) or the deliberate decision not to mention a study in order to give a flavour of originality to one's own work (a serious felony, and, alas, very widespread). One should mention that due to the immense amount of scientific work published, even scientists like me who are very careful about this problem are not immune from mistakes. Today, thanks to powerful search engines, almost nothing escapes a careful and meticulous search. But it may happen that, when facing an overwhelming number of contributions on a particular topic, one may miss something relevant. As I said before, this can happen from inadvertently overlooking or from the impossibility of really knowing all that has been done. The other case, cheating by reproducing already known data without mentioning the source, has become, unfortunately, quite common and represents one of the disgraces of modern science.

Stories of plagiarism

The first time I came across an obvious case of plagiarism was about 10 years ago. A couple of years earlier we had published a rather important work on the topic of defects in titanium dioxide, another common material of great technological importance. A year later, I saw with some surprise a study on the same subject, made with almost identical methodologies, published by a Korean group in a major international journal. Being curious, I immediately read the paper to see what was different from our approach. As I was proceeding with the reading, I found more and more similarities with our published work. Amazement turned into outrage when I came to one of the key figures in the Korean study: it was practically the same as the one in our article. I could not believe what my eyes were seeing. A coincidence? Possible, but very, very unlikely. The suspicion that those authors had duplicated our work in a fraudulent manner was very strong. The worst thing was that there was no mention of our work in the list of bibliographical references. Disgraceful!

In such cases, the risk is that the priority of the results will be lost with time, and the original contribution will be confused, making it impossible to determine who was the first to report that particular result. A reaction was absolutely necessary. First of all, I wrote a rather annoyed email to the authors, saying that I had read their paper with interest, but sent along a copy of ours, an implicit bill of indictment. I waited, in vain, a couple of weeks, but no answer arrived. At that point the suspicion that the case was not accidental became almost certainty, so I took paper and pen (in a metaphoric sense) and I emailed the editor of the journal, enclosing a copy of our paper, clearly showing the date of publication and explicitly mentioning the extraordinary similarity of one of the figures.

Some time later, the editor wrote back, saying that the resemblance was remarkable, and that it was unacceptable that our work was not even quoted. The answer, also addressed to the Korean authors, asked them to publish an *erratum*, a short article where changes or corrections are made to a previously published article. Unfortunately, the erratum never appeared nor was there any reply from the authors, reinforcing the feeling that they had a guilty conscience. I confess that I was really astonished. I was disturbed by the fact that this substantial duplication of an existing work had escaped the review process; I was amazed by the lack of feedback from the authors both to my email and to the Editor's request; finally, I was surprised that two almost identical papers on the same topic could stand in the literature without any reaction. The world of science is changing, I thought. But that was just a beginning.

When, a few years later, I was invited to a well-known university in Sweden as an opponent to a PhD dissertation, the fact that the plagiarism problem was growing considerably in size, starting to stimulate response actions, became clear to me. In many European countries, and certainly in Sweden, the day of the PhD defense is the most important in a person's life, with the same relevance as (if not bigger than) the day of marriage. But playing the role of the opponent of a doctoral thesis in Sweden is a challenging task for another reason as well. The doctoral degree examination involves a doctoral student making a presentation of his or her work, followed by a two-hour discussion in which the opponent must turn the job upside down, in and out, to ensure the student is perfectly aware of what he or she had been doing. All of this takes place before the public and a jury, who, however, silently watch the rude confrontation.

In order not to fall short of arguments or questions (which would cast a shame on the examiner even more than on the student), the opponent must prepare his role with great care and assimilate the student's thesis work in great detail. It is not enough to read it, you really must dissect it, line by line. Well, when I arrived the day before the public defense I had come to the end of this tiring procedure. A dinner was given, attended by the student's tutor and by members of the thesis committee, and I immediately realized that something was wrong. In fact, I learned that a member of the jury had discovered that a chapter of the student's thesis had been copied, word by word, from another doctoral thesis. How the jury member had come to this discovery remained a mystery, but evidently he hit the mark. The student was invited that same evening to correct the thesis, and to open the presentation the following day with an address of apology for the event. A PhD student is quite nervous anyway the day before the exam, and I could not imagine how that poor person would pass that night. The following morning the ceremony began with a short speech from the candidate who apologized for copying part of that controversial and collateral chapter. Lack of command of the English language, and the fact that she was somehow unfamiliar with the subject were the reasons given to justify a serious carelessness. It seemed to me a sufficient humiliation, suffered in front of me, the committee, and the public, including friends and family members.

The presentation followed, and after that, a lively and thorough discussion of the results for nearly two hours, as one expected according to the protocol. When the discussion ended, the committee members convened to formulate the final judgment. I have experienced this step dozens and dozens of times in virtually all European countries: jury members discuss answers, mastery, and presentation skills of the candidate, and formulate a judgment when required. After usually an hour or so, the committee leaves the meeting room and goes back to the lecture theatre for proclamation, applause, kisses, embraces, and sometimes some emotional tears. Instead, at the beginning of the meeting, a spokesman for the university was introduced to us. He was not part of the jury, but nevertheless attended the discussion. It seemed strange to me, surely it had never happened before. The guy remained in religious silence as we discussed, rather unconvincingly, the final judgment to be appointed to the student. Once consensus was reached, as we readied to go back to the lecture theatre and celebrate, the mysterious person qualified himself as an emissary of the university's ethics committee. He informed us that the awarding of the doctor's degree was suspended, and that the case would be transferred to the committee itself for decision. But, I thought, why did he not tell us beforehand, avoiding all the charade? And now, who was going to tell the student and their friends? So, we started a long, animated discussion, initially convinced that it was just a way to scare the candidate a bit, and perhaps to downgrade the final judgment. We were locked in the room for two hours, but the guy was unmovable. No PhD degree, not today. No celebration, no party, everything screwed up. Like a marriage with all the relatives waiting for the bride who does not arrive. The temperature between us dropped to absolute zero. The impression was that the student, positively guilty, was to be sanctioned severely to establish a rule. I realized that in some environments the phenomenon of plagiarism had reached dangerous levels, and that strong actions were beginning to be taken to counteract it. In the follow-up, I learned that the candidate graduated almost a year later.

I met that former student recently, now an Associate Professor, while visiting a prestigious University in China. I was contacted for a short meeting, so we could talk about many things, including position, research interests, and future prospects. But we both chose to skip the story of the PhD defense: such things are much more properly forgotten than revived.

How widespread is plagiarism?

One might wonder to what extent the phenomenon of plagiarism is widespread in the scientific world. The answer is not simple, as well-conceived plagiarism is very difficult to discover. But if plagiarism is raw, and just means copying and pasting an already published text, then there is no escape for the culprits. Powerful software that can compare a new text with huge databases of extant papers is available. This software can trace word sequences or even just characters that duplicate texts already known in the literature. The editorial staff of scientific journals and even many universities currently use this software to avoid having, in articles or doctoral theses, parts of texts, or even whole paragraphs, copied from other sources. Recently, Daniel Citron and Paul Ginspart of Cornell University analysed a set of 757,000 articles published between 1991 and 2012 on arXiv.org, an open-access site where mostly physics articles are reported.[20] Using a special software, they searched for all occurrences of same seven-word sequences in these articles. For quick reference, consider that two articles with 100 overlapping sequences of these seven-word sets may have about thirty-five sentences in common. The study highlighted the extent and geographical distribution of duplication cases. The threshold for considering a work as duplicate was set at having at least 20 per cent of the text in common with another publication. It turned out that less than 1 per cent of the duplicates came from countries like the USA or UK. The countries with the highest frequency of duplications are, in alphabetical order, Bangladesh, Belarus, Bulgaria, Colombia, Cyprus, Egypt, Jordan, Iran, Pakistan—in other words, they are all emerging countries that are just beginning to enter the global world of scientific production, but are essentially in absence of a well-established tradition and well-rooted ethical values of science.

Not only texts, but also images, are affected by the same problem. Analysing over 20,000 biomedical papers, recent studies[21,22] have found that about one paper out of twenty-five (4 per cent) contains duplicate images. The number was about 1 per cent 10–20 years ago. The level of image duplication varies among journals, being, for example, more than 12 per cent in the *International Journal of Oncology*, and only 0.3 per cent in the *Journal of Cell Biology*, which has been routinely scanning images before they are published.

As I said before, however, the problem is much more elusive. Stealing an already published idea and disseminating it again in different words as if it were original is not very difficult for an experienced author with some command of English. If well perpetrated, this kind of plagiarism escapes any control, except perhaps the attention of those who have such a deep knowledge of the field they are able to recognize duplication. But with the mass of existing literature before us, is there anyone who can claim to be in that position? Of course not, and the vetting nets through which corrupt or dishonest authors must pass are very loose. There is little arguing with that.

4

Judges and defendants

At first glance, an email arriving on 16 October 2015 looked quite like any other email of that kind received previously: invitations to submit articles to journals never heard of before, or to give lectures at obscure international congresses in exotic locations. So many and so invasive that in most cases I have gotten used to not opening them and transferring them directly to the trash folder. But this one somehow attracted my attention, perhaps because of its unusual amount of shamelessness. 'Dear Gianfranco Pacchioni', the mail began in a friendly and confidential tone,

> we invite you to consider the journal 'Modern Integrative Molecular Medicine' to submit the results of your research. We greatly appreciate the contribution of our authors to improve the content, the scientific quality and the clinical relevance of our journal. We make every effort to ensure that we only publish high quality papers.

This was more than enough to attract my curiosity and to give me an itch to read on. I am totally alien to the field of integrative medicine and I have never worked in it or on any related topic. I work on completely different things and did not see how I could contribute to the scientific quality and clinical relevance of a medical journal. It seemed to me a terrible start for someone who invites you to publish a scientific article in a new journal. If you address a teetotaler trying to sell him a bottle of wine, the

The Overproduction of Truth. Gianfranco Pacchioni.

DOI: 10.1093/oso/9780198799887.001.0001

least one will conclude is that your marketing division is rather off the mark.

The email continued: 'Our goal is to have our first impact factor soon', which means that they did not yet have one. *Excusatio non petita accusatio manifesta!* (excuse given without being asked is patent self-indictment). As we will see later, this is one of the key points in modern science, a journal's impact factor: without it, a journal is almost nonexistent, and in this case, apparently, the impact factor was not there yet. 'We have maintained and improved our efficiency in the review and editorial process', continued the email, 'so to ensure publication of the work in no more than 7–10 days.' Fantastic! Just one week to decide whether the work can be published when it usually takes months. Another miracle, I thought. 'With your support', proudly ended the email, 'we intend to make this journal one of the most important scientific publications in the world. Also on behalf of the Editorial Board, I invite you to consider 'Modern Integrative Molecular Medicine' as the first choice for the publication of your most important work.' Signed: The Founder, Editor-in-Chief.

So, let's summarize: an invitation addressed to someone having no part whatsoever in the field of integrative molecular medicine, a journal not yet listed in the databases and without an impact factor, and ridiculously short reviewing times for submitted papers; yet, it was an ambitious goal: to make the journal one of the most important scientific publications in the world. Not bad, a clear example of unpretentiousness and good professional practice! Unfortunately, I was confronted with a clear case of a new way of producing scientific knowledge, with a typical example—not the first, not the last—of how the way of communicating science has become corrupted in recent times.

Such emails, unfortunately, are not an exception. Every active, and even no longer active, scholar nowadays has the same daily experience: invitations to publish in unheard of journals, to attend conferences at beautiful but improbable locations in Thailand or the Caribbean, to organize and chair conference sessions with a wide range of topics to choose from. A real menu, a sort of online catalogue of opportunities to acquire some exposure, with little or no relation to the scientific content of your work. But all sharing one prominent feature: sooner or later the bill arrives. A bill—literally, I mean that you are asked to pay money: to pay for publishing, to pay exorbitant registration fees for the congress of your choice, and so on. In short, a market where door-to-door salesmen have been replaced with sales pitches sent by sophisticated electronic communication tools. But how did this happen? To understand this, we need to take a step back and explain the working principles in the validation process of results of scientific work.

First level of judgement

Science is one of the few human activities where the protagonists are at the same time judges and defendants. In the 'normal' world, the roles of those who evaluate and of those who are evaluated are strictly separate. It would not be possible to have a football referee who at the same time plays football on a team, or an external administrative auditor who also acts as a board member of the same company. And in court, nobody would like the idea of swapping from time to time the jury's foreman with the defendant. However, this is what regularly happens in the scientific world. It is called peer review, or peer evaluation, and it is considered the pillar of the self-assessment process on which the entire system rests. It normally takes place free of charge, in the sense that a researcher or scientist sooner or later is called to evaluate the work of colleagues, always on the basis of an implicit and silent principle of reciprocity. It is therefore indispensable to understand how the process works in order to appreciate how vital this step is in the day-to-day activity of those busy with doing science. And also why this process, at present, is proving fragile before the challenges posed by modern science's tumultuous growth.

The peer-review process is a fundamental stage of scientific communication. Apparently, it was applied for the first time over 300 years ago when the British Royal Society started the practice of publishing scientific papers in its *Philosophical Transactions*. The current form of peer review is about 40–50 years old. In older times the selection was made by editorial committees or directly by journal directors, the editors. *Nature* introduced the peer-review process in 1966 when John Maddox became its legendary Editor-in Chief. In contrast, the famous *Proceedings of the National Academy of Sciences* introduced peer review only a few years ago. Peer review is the main mechanism for deciding whether a paper is valuable and therefore deserves to be published. But it has also become the primary mechanism for deciding whether a scientific project is robust, innovative, and realistic, and can therefore be funded, or whether the scientific profile of a researcher is appropriate for an academic position. In practice, the whole process of certification, selection, and funding of research goes through one or more peer-review stages. But what, precisely, is peer review?

Increasing specialization makes it more and more difficult to evaluate scientific undertakings or projects outside one's field of interest. So, a primary task for scientific publishers or funding agencies is to find experts that can understand the contents of a new document or of a new project. Accordingly, these operators turn to other researchers engaged in the same field who may have published reasonably recent work. Each scientist then may be acting at the same time as evaluator, author, and project principal investigator: hence the 'peer review' denomination. Any submitted paper or project is sent to one or more referees (the jury members) who express their judgment anonymously and independently. Reviews may be in agreement or may give rise to disparate statements that the authors (the defendants) have the right to appeal in their replies, a kind of plea in which one tries to better explain one's point or provide a rebuttal to the criticisms. The final decision is in the hands of the editorial or scientific committees, who are committed to taking into account the opinions provided by the judges, the referees. Each of us, then, at alternating times, puts on one or another hat, resulting in an intense flow of documents travelling over the invisible wires of the Internet. The goal is to provide scientific progress with those aspects of rigor and trust that are the fundamentals of the advancement of knowledge in the field of natural sciences.

Like every complex system, peer review is not without defects. Referees are flesh-and-bone men and women, and as such they may be wrong at times. Generally, in good faith even if judgement in a number of cases can be altered by competing biases, ignorance or simple frustration cannot be excluded. All facets of the human character are in each of us, and as such are unavoidable, even though these devious aspects can be mitigated by comparing different independent evaluations.

There are also aspects of 'psychological subjection', though no one ever admits them. A paper coming from a celebrated, internationally renowned group may prompt a more permissive attitude in the evaluators, especially when they are young or less experienced researchers. And the opposite is also true: a work that comes from an obscure research institution in an emerging country is likely to be read with a more critical eye and a bit of skepticism, even more so when it comes to the funding of research projects. Unfortunately, it is almost inevitable that a research proposal from Cambridge appears more credible to a reviewer than one from an unknown Indian (or I could say Italian) university. So it has been proposed to revise the process whereby the referee is unknown to the author and vice versa; that is, the identity of the author or proposer is unknown to the evaluator. It is called double blind, in which a text or project is evaluated without knowing who produced it. Adopted by some journals, this procedure has some limitations as well, because it is quite possible that the identity of the authors can be guessed from the nature of the proposed work, or from the list of citations, or from many other hints. And in any case, the size and composition of the research team carries an important weight when evaluating the prospects and the reliability of a project.

Despite its many weak points, no substitute for the peer-review procedure is currently in sight. According to the majority of active scientists, peer reviewing remains the best way to validate scientific progress.[23]

Appeals

If all this seems clear and straightforward in its conceptual framework, practical application can be far more difficult. For some years I have been a member of the Editorial Board of *Physical Review Letters*, one of the most prestigious physics journals in the world. One of my tasks has been to evaluate the authors' appeals, an extraordinary and uncommon procedure among scientific journals, but which is a further guarantee of the fairness of the whole evaluation process. Here's how it works: when authors submit a paper, the manuscript is sent to at least two experts who formulate their opinion. If these judgements do not provide convincing motivation for approving publication, the work is dismissed and returned to the authors. However, if there is room for a rebuttal, the authors can modify and resubmit the manuscript that is then sent again to the same reviewers plus, usually, to a couple of other experts. Again, if none of these evaluators express criticisms or concerns, the paper is cleared for acceptance. The difficulty in publishing in a prestigious journal is also the reason everyone wants to see their paper appear there: climbing a steeper mountain gives a lot more satisfaction and prestige than climbing a mild hillside. Let's keep in mind that, as we shall see, the reputation of the journal where a paper is published has become extremely important for a scientist's success; thus, the authors' commitment to convince the referees to publish their work becomes obvious.

In most cases, however, the paper gets rejected even after giving authors an opportunity to replicate. At this point, the normal review process is over, and this is where I enter the game. If the author is particularly convinced of the study's validity and believes that the review process and the final decision have been unfair, he or she can appeal, providing solid motivations. In this case, the voluminous file that contains the original manuscript and all correspondence, reports, judgements, and counter-judgements is sent to a member of the Editorial Board who is empowered to formulate a further evaluation, this time indeed final. While an article in *Physical Review Letters* cannot exceed four printed pages, the appeal is usually 40–50 pages long; the most extensive appeal I ever received was 90 pages altogether. This means that in order to convince the reviewers of the robustness of their results, summarized in four pages, the authors end up writing at least four to five times as much text, in what is an exercise in oration (even if in writing).

These heavy files provide an interesting view into the world of science. They show the unshakable trust authors place in their work, to the point of defending it tooth and nail against any raised criticism. They also demonstrate how each of us firmly believes that the field in which we work is without doubt the most important of all. In some cases, I'm sorry to say, there is a good amount of pretentiousness and arrogance, which does not fit well with the poise of a scientist. With few exceptions, each point is made in a proper tone, and the argument is deployed with rigor and determination, but from time to time rude or offensive phrasing may appear. The final decision to accept or to reject is often difficult, subtle, but not without misgivings. One recalls that behind publication, the ultimate stage in the production of knowledge, are concealed years of hard work, along with expectations, competition, ambitions, career prospects, and so on. Nevertheless, the role of judge of last resort must be independent of all this and must only focus on the quality, innovation, and impact of the proposed work. It is a long, complex, expensive process, but it guarantees the best selection, and contributes decisively to the prestige of the journal. A process that, above all, requires a precious commodity: time.

This is a particularly virtuous case of the review process. All major journals adopt similar criteria. But with the continuous growth in results, new problems that were inconceivable until a few years ago have arisen. Just to give an idea, when I started my career the flow of incoming submissions to a journal's editorial office was on the order of a few manuscripts per day. The editor personally chose a couple of reviewers for each paper, thus starting the evaluation process. Today, every journal of some reputation receives tens, or even hundreds, of submissions per day. Such a mass of incoming manuscripts requires a complex office apparatus to collect the submissions and decide, devoting on average 20–30 minutes to each one, whether they stand a chance of being accepted through the peer review process. If the submission deals with a topic which does not fall within the scope or readership of the journal or if, in the selection staff's opinion, there is no chance the paper will successfully pass through peer review, it is rejected outright. In some journals 80–90 per cent of submissions do not get beyond this stage. This figure is quite understandable when one considers the number of submissions. It would be simply impossible to send all these papers to external experts, with unmanageable handling and organizational work.

That is why in many journals, certainly in all prestigious ones, the first selection is done by the staff of publishing houses, professional editors who are no longer involved in active research. In this respect, one cannot speak of a true 'peer' review, rather of a preliminary selection whose goal is to ensure that the journal publishes only the 'best' research work: in more plain words, only those papers with potential or promise to generate a high return in terms of visibility and reputation for the journal itself.

There are well-established methods, to be discussed later, for assessing quantitatively the importance of a scientific journal, methods that somehow determine its popularity and prestige. The machinery is not too different from that used by newspapers to scoop and select the news to be published or highlighted on the front page.

How many scientific journals?

What happens to those papers that do not pass these preliminary filters or are rejected after being peer reviewed? Nothing particularly dramatic. Their authors send them to a less prestigious, less demanding journal whose selection criteria are a bit looser. I can assure you that in this sort of downgrading process, today everything is possible. After the top-level journals, there are, in fact, many journals of medium-high, medium, low, and very low quality. This has to do with the ever-growing number of 'scientific' journals being published today. In 1964 the influential database Science Citation Index listed a total of 600 peer-reviewed, scientific journals. By 2004, 40 years later, the number had increased to 5,969, or ten times higher, according to a study of 2010.[24] However, not all scientific journals are listed by the Science Citation Index, and in fact, in 1996 the actual estimate was about 11,000 academic journals.[25] They had become 14,694 in 2001.[26] According to a report by the International Association of Scientific, Technical and Medical Publishers,[27] in 2014 there were approximately 28,100 active journals featuring a peer review process, plus a further 6,450 non-English-language journals. The number of journals has thus grown steadily at a rate of about 3.5 per cent per year (see endnote 26). And this is without considering conference proceedings, a parallel, widespread form of communication in areas such as information technology and engineering.

So, with a bit of perseverance, it is not too difficult to publish an article, thanks to not very committed editorial committees, to a not particularly careful or just casual selection of reviewers, and to very permissive acceptance criteria. Of course, a paper published in one of these journals will have less impact and resonance, but it will find its place in the researcher's CV, inflating its bulk. And even if you could not find a complying journal for your paper, there is always a way out: just pay, for example, by participating for a conspicuous fee in one of the myriad of congresses organized by specialized, purely commercial agencies in splendid tourist locations, and presenting a contribution to be published in their proceedings. Or simply use the emerging phenomenon of open-access journals where alongside good and valuable stuff one can find pure and simple scientific junk.

Open access

Open access is a recent phenomenon, entirely related to the dematerialization of communication. Up until the advent of the Internet, say, 15–20 years ago, all journals were in printed-paper form. Single issues came at regular intervals to libraries where they were periodically bound into large, heavy hardcover volumes that quickly filled library shelves that were less and less adequate to accommodate them. With the advent of electronic communication, all this gradually disappeared. Print journals are published alongside electronic, online versions, and in most cases the digital versions have completely replaced the paper ones. This has paved the way for a completely different, previously unknown kind of publishing, the impact of which is still to be fully understood: I refer to open-access journals. This concept entered the scientific literature in the early 2000s.

The idea is that the results of research work, generally produced using public money, should be made available to everyone without having to pay the high subscription fees required by traditional publishers. Quite often, subscription fees are really high (up to something like several thousand euros a year). Traditional publishers of scientific papers retain a strict copyright, allowing access only under subscription or by payment of a fee. Open access is a noble concept born to counteract this policy: the costs associated with publication, that is, review of texts, editorial management, peer review, website maintenance, etc., are paid directly by authors at the time the paper is accepted for publication. Under this agreement, copyright is not transferred to the publisher and published articles can be made available without delay. So, if you search the Internet to find out the latest news on asthma pathology, you will find some scientific articles that you can only access as far as title and abstract without paying. These are published by traditional journals and are copyrighted. You may find others whose content is entirely accessible for free: these are open-access articles. Authors must have paid a sum, typically between 1,000 and 2,000 euros, to have the results of their research published.

When the first open-access journals appeared, they was thought to be the beginning of a deep, positive revolution in the mode of scientific communication. Unfortunately, a few years later, the scenario was complicated and confusing, and far from positive. In October 2017 there were 10,114 open-access journals based in 122 countries, an increase of nearly 50 per cent over the previous six years,[28] for a total of 2.6 million articles. This is explained by the relative ease with which this type of enterprise can be created. Everything is dematerialized, everything travels on the Internet, and there are no printing, shipping, distribution, or subscription costs. To become an open-access publisher, all you need is an editorial office with just a few people, located at some point on the terrestrial globe, and a website. Of course, you also must muster together a respectable editorial committee to gain credibility, but with millions of active researchers around the world, all looking for a pinch of exposure, this will certainly work out easily. Many, in good faith, agree to be members of these committees, providing a look of scientific steadiness and a modicum of respect to the journal. As a result, a multitude of open-access journals (not all!) now qualify as one of the most serious risks to the integrity and credibility of the scientific communication system. Many of these journals, created from scratch by unscrupulous publishers only interested in profit, have adopted very permissive publishing policies whereby the peer-review process is kept to a minimum, or is completely absent. All this has gone so far that the term 'predatory journal' has been coined.[29]

Online piracy

A survey conducted in 2013 for the American journal *Science* by John Bohannon, a Harvard biologist and journalist, gives statistical significance to the problem of misuse of open access. Encouraged by endless invitations to publish in dubious quality journals, Bohannon decided one day to 'see', just like in a poker game.[30] Between January and April 2013, he submitted hundreds of copies of a counterfeit paper in which he reported the discovery of a miracle drug, extracted from a lichen, capable of fighting some types of cancer. The bogus paper had been written in such a way as to be definitely rejected in any respectable peer-review process, due to a series of planted inconsistencies, errors, and even preposterous suggestions such as bypassing the clinical trial phase, an indispensable aspect for any serious work in drug research. In short, Bohannon committed himself to a paper with so many flaws and contradictions that no serious reviewer would have failed to detect them. To nail in the message, he used as author's name an imaginary African researcher working in a non-existent institution in Eritrea.

With the help of a software program, he generated hundreds of slightly different versions of the paper and submitted them to 304 open-access journals. The outcome turned out to be shocking. About half of these journals, exactly 158, agreed to accept the paper without rising an eyebrow, while only 98 rejected it. Of the remaining 48 journals, 29 no longer existed just a few months after their creation, with their websites abandoned. As clearly appeared, of the 255 cases in which there was a positive or negative response concerning publication, about 60 per cent never went through a peer-review process. Only 36 submissions out of 304 produced reviews from referees to whom the (dummy) author was called to respond.

Not a bad business

Obviously, one cannot and should not generalize, and not all open-access journals are of low quality. On the contrary, some are very well-established and trustworthy. What is certain is that this is a great business for publishers. *Nature* launched an open-access offspring in 2011, called *Scientific Reports*. Obviously, this is a serious journal with a robust peer-review process, but figuring out what kind of business is concealed behind this model is relatively straightforward. In 2016, *Scientific Reports* published 21,045 articles, a 190 per cent increase over the previous year. The publication cost is €1,165 per item, which for the articles published in 2016 makes a revenue of over 24 million euros. Still nothing compared with *PLOS One*, a well-known open-access journal that was among the first to appear ('PLOS' is an acronym from Public Library of Science). Born in 2006, *PLOS One* published 31,236 articles in 2015, at a cost of $1,495 per item, for a total revenue of $46 million—and it is not alone: there are several other more specialized titles in the family, *PLOS Biology*, *PLOS Medicine*, etc. Not bad, a really good business. And without printing a single page on paper!

But how to separate the wheat from the chaff? How can you tell whether a given journal is reliable? In 2008 Jeffrey Beall, a Colorado University librarian, began writing a sort of black list of scientific magazines suspect of being 'predatory' journals.[31] It was a kind of reverse Michelin Guide, listing restaurants where no one would want to eat. This list has grown over time and was updated continuously. By October 2016 the list had more than 1,200 entries. In 2017, however, Beall decided to close the website due to increasing pressure and threats from open-access journal editors and from academic authorities, who went so far as to put his job in danger.[32]

It would not be proper to criticize the rapid growth of open-access journals without pointing out that, to stick with business issues, we are witnessing a general proliferation of journals, even those traditionally distributed by subscription and subject to copyright. Publishers are the first culprits, without distinction in this case between dedicated houses, be they of private capital, such as Wiley, Springer-Nature, and Elsevier, or stemming from scientific societies that publish specialized journals and whose primary purpose has been—at least, until recently—cultural advancement rather than money making. Today, all publishers are unanimously committed to producing every year new periodicals, new series, new sub-sectors, and increasingly specialized journals. As a result, subscription costs have swelled along with the revenue. However, it seems that the hunger for publications is such that as soon as these new journals are created, they immediately attract a significant number of submissions, in turn stimulating publishers to invent more journals, perhaps just to cover a research subsector before a competitor does.

All this encourages the proliferation of low-quality scientific undertakings. But watch out: we have not yet reached the bottom of the downward ladder. A new phenomenon is dawning on the scientific horizon: the selling and buying of scientific works.

Publications at the supermarket?

I think it proper to point out that, although I have had some direct experience with all the other problems and distortions the world of science is enduring today, this is not the case for what I'm about to describe. I have only found traces of it in articles that have appeared in mainstream journals like *Nature* and *Science*. In particular, *Science* dealt with the market of scientific papers in a 2013 investigation by Mara Hvistendahl, who demonstrated how acute this problem has become in China.[33] A five-month investigation revealed a system that conceals a flourishing black market of publications, involving shadow agencies, corrupt researchers, and compromising publishers. The subject of the business: scientific papers that are included in important databases such as the already-mentioned Science Citation Index. *Science* documented that for a sum ranging from $1,600 to $26,300, you can become co-author of a paper. In some cases the price can be very high, approaching the order of magnitude of the annual salary of a professor in China, but the return may be equally profitable, as the number of published papers and their placement in scientific databases play a key role in the career of young Chinese scientists. In fact, today, for a young researcher in China, publishing in a journal included in the Science Citation Index and possibly with a good impact factor can pave the way for an academic career.

But how does the system work? Apparently, you can buy a manuscript from an online catalogue, simply and smoothly, often with guarantee of publication. *Science* had spotted 27 agencies that sell articles to appear in the Science Citation Index. In particular, using the Chinese search engine *Baidu*, an ersatz Google in a country in which Google is banned, all that is needed is to type a few keywords (in Chinese) to find dozens of agencies offering that service. This phenomenon is the consequence of the incredible growth that Chinese science has seen over the past decade. The number of scientific papers published in China has risen from 40,000 in 1999 to 400,000 in 2013, a 1,000 per cent increase (see endnote 6)! It is impossible to ascertain how many of these papers have been produced using methods such as those described above, but, according to *Science* their number is quite high.

Another kind of misbehaviour unheard of until recently is the addition of a name to the author list once the paper is accepted. This practice is not widespread and is justified by the fact that sometimes, in the reviewing process, referees require the performance of additional experiments. As a rule, these are carried out by the original research group, but in special cases it may be necessary to apply some expertise that the group does not possess. New measurements are therefore made by researchers who have not been involved in the study, thus the motivation to add their names at a latter stage of the process. This is an uncommon occurrence, one I have never met in my professional life, but that cannot be ruled out a priori. In such cases, however, the contribution of new authors must be adequately specified and motivated. Apparently, this procedure is quite common in some less conscientious journals, giving way to the suspicion that authors are added, once the paper is going to be accepted, upon payment of a fee.

Another clearly unethical behavior, in connection with the peer-review process, has emerged recently. It is common practice that when an author submits a paper to a journal he or she also provides a list of 4–5 potential referees. The reason is simple. Who better than the author knows the field and can recommend experts capable of providing a sound assessment of the paper, making useful suggestions on how to improve it, and providing a robust validation of the study? This assumes that the goal is to contribute to augment knowledge. But if the purpose is just to get another paper published, the review process is only a hindrance, for it slows down publication and can result in rejection. Recently, the journal *Tumor Biology* found that reviews submitted in support of 107 papers had been fabricated and retracted the papers.[34] The reason was fraud in peer review. Researchers, or companies acting on their behalf, recommend scientists as potential referees, but the email supplied for the reviewers route back to the authors themselves or to accomplices, who then write positive reports just to get the paper published. Of the more than 100 papers retracted, it was found that 9 were fraudulent, and 12 had been purchased from third parties by the supposed authors (see endnote 34).

All the above is certainly marginal and uncommon, but the very fact that we must talk about these problems, and that they do occur, is a clear sign of the deterioration of the ethical standard on which scientific progress has been based. We are witnessing the transformation of scientific enterprise from the dedication and vocation of a minority of inspired people into a vast and hungry market, with all the detrimental aspects that this entails. As much as this paper market is to blame, it still is a restricted phenomenon. There is another aspect of much wider diffusion and negative impact that, although ethically less problematic, can have devastating consequences. I refer to the spread of studies that are irrelevant, or wrong, or plainly irreproducible.

Ferroelectrics go bananas

The process of detecting and correcting obvious errors in published works is complex and not without inconveniences. Recently, a group of nutrition and obesity researchers spotted and pointed out a number of studies affected by obvious statistical errors that inevitably led to wrong conclusions.[35] After trying to rectify a dozen or so cases of evident inconsistencies by sending letters to journals or commentaries to editors, our heroes threw in the sponge due to too many hurdles: long delays in getting answers, journal editors reluctant to deal with this kind of problem, and, in extreme cases, journals asking for a fee to post comments or to withdraw erroneous papers. Healthy science requires a high-performing, error-correction mechanism, as Allison et al. aptly pointed out in their 2016 *Nature* article (see endnote 35).

Another interesting and funny case is an ironic article by James F. Scott, one of the top experts in ferroelectric materials, published a few years ago.[36] We use ferroelectric materials in many electronic devices, and they have a strategic importance in many technologies. Such materials are characterized by their typical response curve to an external electric field, called a hysteresis curve, as shown in Figure 3d.

Experimentally, in order to understand whether a new material is also a ferroelectric, an electric field is applied and the response is plotted in a graph. The material is ferroelectric if this shows a hysteresis cycle. Unfortunately, many substances under the same test give rise to similar curves, such as that shown in Figure 3a, yet with significant differences from the true thing. While non-experts can be deceived, a specialist must be able to distinguish a true hysteresis curve from a fake one. Indeed, the problem is well described in many basic textbooks. Prompted by the growing number of papers reporting cases of new alleged ferroelectric materials, Scott published a brilliant

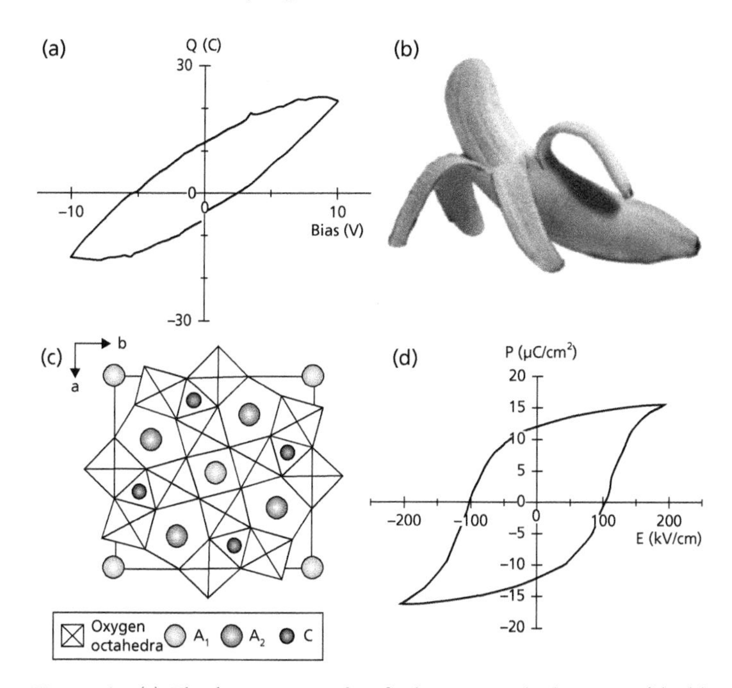

Figure 3 (a) The hysteresis cycle of a banana peel, shown in (b). (c) The structure of a real ferroelectric material. (d) The corresponding hysteresis curve.

Source: From J. F. Scott, Ferroelectrics go bananas, *Journal of Physics: Condensed Matter* **20**(2), 2007.

article entitled 'Ferroelectrics go bananas', where he shows that even a banana (yes, the tropical fruit) if subject to an external electric field exhibits a polarization curve, as shown in Figure 3a, without obviously being a ferroelectric material. Scott explicitly quotes a dozen clearly off-the-mark papers that appeared in the literature, and claims to be aware of at least a hundred similar cases. If your hysteresis cycle looks like that of a banana, Scott ironically concludes, please refrain from publishing your results!

Irreproducible results

The issue of reliability of scientific results goes hand in hand with the equally if not more important issue of reproducibility. In their rush to publish and deliver outstanding results, researchers are often pushed to evaluate performances, yields, and effects in exceedingly optimistic ways. An example of this has been discussed recently, and is related to the efficiency of next-generation solar cells.[37] To be economically attractive, a solar cell must have at least 20 per cent efficiency in the transformation of solar radiation into electric energy. In 2010, a new type of cell, called 'perovskite cells', were discovered with efficiencies purported to be on the order of 20 per cent. Unfortunately, there was some skepticism and there were problems of reproducibility with many of the supposed performances reported in the literature (see endnote 37). All this is due to an intrinsic problem of poor stability of these substances in open air and in wet environments, which makes efficiency measurements complex and scarcely reliable. Trying to cope with the overwhelming excess of optimism, some journals have begun to require that the declared efficiency be verified by independent, certified laboratories.

Lack of reproducibility of results poses one of the outstanding dangers to which contemporary science is exposed. What is at stake here is the mainstay on which the entire system rests: a result is considered solid, trustworthy, and reliable only when it can be independently reproduced by others. Needless to say, the different conditions under which the experiments are carried out, the complexity of many of the experiments, the need to employ very sophisticated instruments not easy to handle, and so on, are all valid reasons why some results are difficult to reproduce, even though they are correct. This is also why, as we shall see, scientific fraud is so difficult to spot.

Lack of reproducibility is more and more responsible for very high economic and social costs. The problem is particularly acute in the field of clinical trials, where new procedures or new drugs are tested for the treatment of a given pathology. According to some studies, 80 per cent of clinical trials cannot be reproduced simply because the inherent statistical analysis either is inaccurate or has been carried out without appropriate care. The Biotec Amgen Company has teamed up 100 operators to try to reproduce 53 fundamental articles in cancer research published in prestige journals.[38] Only six of these, or 9 per cent, have been reproduced. Likewise, Bayer has carried out tests to verify 67 studies in the field of oncology, female health, and cardiovascular diseases. The results were reproduced only in 14 cases out 67, or about 21 per cent (see endnote 38).[39] According to another study published in *PLOS Biology*, $28 billion are spent each year in the United States in basic biomedical research, but many of the produced results cannot be replicated.[40]

The reasons are multiform: improperly designed experiments, use of impure reagents or poor materials, shaky analysis of data. It is estimated that the rate of non-reproducibility in this field is in excess of 50 per cent, consequently causing colossal economic damage. But the problem is not only with clinical trials, as shown by a survey conducted in 2016 by *Nature*.[41] More than 70 per cent of the 1,576 researchers of various expertise who filled in a questionnaire on the topic stated that they were unable to reproduce results from other groups, and more than half of them even failed to reproduce their own results! Despite this, most respondents still expressed confidence in the validity of the scientific literature. According to the survey, the reasons behind low reproducibility are pressure to publish and tendency to report partial results.

What transpires from all this is that one cannot assume that a result published in a scientific journal is also necessarily rigorous, reliable, relevant, and reproducible, as one would expect. All of these features are crucial for the advancement of scientific knowledge but they are not always guaranteed by the peer-review process described at the beginning of this chapter. Even when an article has gone through careful review, more and more often we witness the publication of erroneous, inconsistent works lacking indispensable foundations. This is the unfortunate and unavoidable consequence of the inordinate growth of the scientific community: a process that parallels a decline in overall quality. In a crowded world of mediocre scientists, it is not surprising that mediocre, irrelevant, and even wrong studies find their way to academic consecration by publication in a 'scientific journal'.

What defense is at our disposal, then? How can one ensure the reliability of any published result? Some quality parameters that can guarantee the relevance of a given scientific product are definitely desirable. With more papers written but with less and less time to read, the need is felt for a preliminary assessment of a scientific result's degree of reliability, even before its specific contents are assessed. Here's where the reputation of authors, institutions, and journals come into play, along with procedures that can somehow capture a reputation and evolve it into a numerical value. Thus, we are led to the subject of bibliometric indices, of rankings and classifications: the mixed blessings of modern science. This is the jungle we are about to enter: the science of measuring science.

5

Units of measurement

Baltimore, 14 March 2016, at 19:30, Hilton Hotel, site of the annual meeting of the Editorial Board of *Physical Review Letters*. About 40 people attend the meeting, preceded by a frugal dinner followed by a lively discussion. London, 7 May 2016, at 10:30 a.m., Grange Strathmore Hotel. Same scenario, only the journal changes; this time it is the *Journal of Physics: Condensed Matter*. I am attending: different meetings, different continents, same discussion: how to improve the performances of the two journals, and in particular how to cope with the fierce competition slowly undermining the foundations on which these are based: the number of submitted articles. In a world where all indicators show a steady increase in published academic work, these two historic journals from two equally highly respected scientific organizations, the American Physical Society and the British Institute of Physics, respectively, are facing the same problem, that is, how to attract more submissions and how to increase their impact. It is not a matter of prestige, not at all. In the long run, it may turn into a problem of survival. In the current, increasingly aggressive and competitive market of scientific publishing, not adapting and not finding appropriate corrections may lead to succumbing to the strain. And this can be the case also for historical journals considered at the top in their field.

The year 2015 marked the 350th anniversary of the birth of scientific journals. In 1665 the *Journal des Sçavans* in France (later called *Journal des Savants*) and the *Philosophical Transactions* of the

The Overproduction of Truth. Gianfranco Pacchioni.

DOI: 10.1093/oso/9780198799887.001.0001

Royal Society in England were founded in order to foster the exchange of information among the few dealing with scientific research on an individual and personal basis, avoiding duplication of results and laying the foundations for the unambiguous prioritization of a discovery. With these periodical publications, the process of academic communication, which until then had relied on personal correspondence with letters and manuscripts, meetings at scientific societies, and publication of books, changed radically. Today, three and a half centuries later, the global market for scientific and technological publications (including journals, books, technical reports, databases, etc.) is valued at over $25 billion (2013 data; see endnote 27). Journals produce a turnover of around $10 billion, while books account for about $4 billion. These revenues come from library subscriptions (68–75 per cent), followed by corporate subscriptions (15–17 per cent), advertising (4 per cent), personal subscriptions (3 per cent), and publishing costs directly covered by authors (3 per cent). It is a flourishing and expanding industry that employs about 110,000 people globally, along with 20–30,000 more freelance journalists, external publishers, consultants, etc.[42] The Dutch firm Elsevier is the largest publishing house in the world, followed by Springer Nature and Wiley. Books are steadily declining, especially printed ones, and the appearance of e-books does not seem to fill the gap. In the above described scenario, open-access journals mentioned in the previous chapter are steadily growing, so that by 2013 10 per cent of published articles appeared in open-access form.[43] The open-access segment of the market continues to grow much faster than the market as a whole but remains small in revenue terms.

To some extent, the scientific publishing market is not so different from that of the television industry, with steady proliferation, in the former of new journals, and in the latter of new TV programmes. In both cases, the products are offered to the public, and their success and level of appreciation are measured through the well-known tool of the television audience, and of what is known as impact factor in scientific journals. Here we come to the heart of the problem, the question that affects the process of publishing present-day scientific results. I am referring to the rush to publish in journals with a high impact factor. Just as for a showman a programme that gets a 20 per cent audience share is more rewarding (even economically) than one that collects only 5 per cent, so for a scientist, publishing in a journal with impact factor 20 is much more prestigious than having their paper appear in a journal whose impact factors is no more than 5.

Measuring the audience of scientists

So, we have discovered that even scientists have an audience measure; it is called impact factor. But what is it? And when was it introduced? Many believe this to be a recent invention. They are wrong. The first person to realize that in a world where science had progressively increased its importance and size so much it would be useful to introduce a numerical quality index of scientific journals was Eugene Garfield (1925–2017), an American linguist and businessman. In the 1950s, Garfield devised the concept of impact factor based on the estimate of the citations elicited by a given publication.[44] That is the pillar, the key to bibliometric science: the citation. How do you establish whether a given scientific work is relevant, whether it has any impact on the community and possibly on society?

Suppose you want to know to what extent a piece of music is popular and how much people like it. The answer is simple. In the past you would count the number of records sold; today, you count how many times the video has been viewed on the Internet, how many times the track has been downloaded, or the number of 'I like' on social networks. For a scientific publication, it is about the same, except that what counts are citations. Every scientific paper refers to previous work. If a paper is cited often, it means that it is remarkable, it has been read by many researchers, and therefore it has left a major footprint. If it is never mentioned, it usually means that it is irrelevant. Of course, there are important exceptions to these rules. There are contributions whose importance is only recognized many years after their publication.

In my field a classic case is that of Pierre Hohenberg and Walter Kohn, who proposed a method for determining the electronic structure of solids and molecules (known as Density Functional Theory).[45] The fundamental paper appeared in 1964 in *Physical Review*. According to Google Scholar, in the two years following the publication that article received only a single citation. Due to the article's complexity, it took a long time for the community to appreciate the importance of that work. Today, half a century after its publication, the paper has been cited more than 30,000 times, but, more importantly, it earned Walter Kohn the Nobel Prize in 1998! Citations, therefore, represent an important parameter for measuring the relevance of a given article, although in certain (rare) cases these may come late. As is often the case, by reducing the evaluation of a complex system to a simple number, one cannot expect to capture all the implications of the system one wants to measure. Classifying a mythic folksinger like Bob Dylan only on the basis of the number of records sold is obviously reductive. But one must start somewhere.

Let's go back to the impact factor and how it is defined. It is simply the ratio between the number of citations received by articles published by a given journal in a given year, divided by the total number of articles published in the two preceding years. Thus, if the magazine *Living with a Cat* had published 1,000 articles in 2015 and 2016 and received a total of 4,000 citations in 2017, the impact factor for that year would be 4, which is also the average number of citations received for each item over that time stretch. If the magazine *Living with a Dog,* which also publishes 1,000 articles, receives, say, 12,000 citations in the same period, its impact factor is 12, or three times higher. We can infer that people like better having a dog than having a cat. This simple concept was immediately translated into a quality index of scientific journals: the more citations received, the more important the published articles, the more prestigious the journal. Things, as everybody can see, are rather linear and respond to a strict logic. Unfortunately, we will soon discover some adverse consequences of all this. Garfield was well aware of the possible problems involved in using the impact factor, to the point where, in an article published in 1999, he wrote, 'I first mentioned the idea of an impact factor in 1955. At that time it did not occur to me that it would one day become the subject of widespread controversy. Like nuclear energy, the impact factor has become a mixed blessing. I expected that it would be used constructively while recognizing that in the wrong hands it might be abused.'[46]

Garfield referred to the fact that a quarter-century after its introduction (the impact factor started to be commonly used around 1975), the index was viewed critically by some, but also with great enthusiasm by many others, to the point of becoming the main parameter for assessing a journal's reputation, but not without some contraindications. For example, many researchers, especially the younger, tend to make a literal and uncritical use of the impact factor, so that a journal with impact factor 5 is considered better than one that does not go beyond 4, when actually the two journals may well be equivalent or it may be that the one with a lower impact factor is better in terms of quality, editorial staff, history, number of readers, and so on. Indeed, this index can be influenced and manipulated in various ways.

Improving the audience

A widespread mechanism for increasing a journal's impact factor is to publish review articles. By providing a more general and wider description of a given subject, these articles tend to be cited more often than papers reporting original results, thus efficiently increasing the journal's impact factor. Besides, impact factors can vary widely for reasons that have little to do with the quality of the publication. For example, in some areas, like medicine, the number of active scientists is higher than that in other fields; thus, the average number of citations per article is higher, directly affecting the impact factor. Sometimes a single article may be enough to artificially alter a journal's impact factor. A classic example is *Acta Crystallographica, section A*. Its impact factor, which has always been around 2, jumped to 49 in 2010 and to 54 in 2011 thanks to an article published in 2008 by George Sheldrick, 'A Short History of SHELX'.[47] The paper reported on the development of a software widely used by the crystallographic community, and invited those using the software to make explicit reference to the paper for the inherent, mandatory citation. As a result of how impact factor is calculated, by 2012 and in the following years the journal's impact factor returned to its usual value of around 2.

All this may sound somewhat technical and rather boring (it definitely is, indeed). However, the career and reputation of individual researchers, as well as the reputation of the institutions where they work, and sometimes even of entire countries, are strongly linked to the impact factor of published materials. On the one hand, journal editorial staff look with unconcealed satisfaction at the increase of their impact factor; on the other hand, department directors, deans, presidents, university administrators, etc., are delighted by the high-impact publications of their affiliates and collaborators. The reason is obvious. A university dean does not have time to read the publications of his researchers; however, he or she can appreciate that a number of these are published in high impact factor journals. Someone maliciously stated once that those people cannot read but can only count, a provocative statement that does not take reality into account, since nowadays no one, not even with the best of good will, can read (and understand) the annual, monthly or even weekly scientific production of any academic institution.

At the top of the impact factor scale are medical journals such as the *New England Journal of Medicine* (55) and *Lancet* (45), followed by the famous *Nature* (41) and *Science* (33). There is then a series of journals with less diffusion and impact, but still with a strong reputation. At the bottom there are hundreds, thousands of journals with ridiculously low impact factors, and sometimes are not endowed with this fundamental index. All this is neither useless nor wrong. It is helpful to know whether a scientific journal publishes, in general, mediocre papers. If I had an interesting result, I would definitely avoid submitting it to that kind of journal. It is a fact that publishing in journals with high impact factor is much more difficult and that you face a tough selection. This gives the reader greater assurance that the work will be of some importance. In a landscape where new results emerge daily, knowing where to find the best ones is vital.

Use and abuse

There is an even more perverse use of the impact factor, and this is when it is invoked to evaluate an individual's performance. In this case, new problems arise. The operation is, in principle, very straightforward. Just add the impact factors of the journals where an author has published his or her work, and you get a simple number. The higher the numerical value, the more relevant the results obtained by that researcher. However, the impact factor, as repeatedly mentioned, captures the reputation of the journal, not the value of a given author's individual work. Unfortunately, the direct consequence of this practice is that young researchers, obviously worried about their careers, are very keen on the impact factor of the journals in which to publish their findings and thus are obsessed about doing so in a prestigious journal. In countries where the academic system is growing very fast, such as in China, a few publications in major journals, as measured by the impact factor, are enough to ensure a good position and the leadership of a research group.

Publishers are not entirely innocent in this ongoing game. I refer to attempts made by many journals to limit the number of published articles by making a selection based not so much on the work's scientific value, but rather on its potential to attract many citations. Exactly in the same way, the choice between two TV shows is usually not in favour of the one with higher cultural content, but of the one more likely to attract a wider audience. All this can be as execrable as you like, but it responds to a steel-strong criterion, in a free-market system. The danger is that scientific progress and its future evolution may end up being directed by market dynamics that should have nothing to do with science. However, science has a strength and an advantage that other human activities lack. If a piece of work is really useful and relevant, it does not matter where it is published. Sooner or later it will be discovered and appreciated. It's just a matter of time.

Fashion, emulation, homologation

Even in science there are trends and fashions. Research topics and themes arise and evolve with tight, sometimes infernal rhythm, maniacal cycles, and a good part of the community jumps on the latest bandwagons in an attempt to create a profile, an identity. A new discovery immediately stimulates huge attention and interest, and thousands, tens of thousands, sometimes hundreds of thousands of scientists jump onto the same theme. By increasing the number of people working on that topic, the related activity increases and inevitably the citations of studies in that field blow up in a kind of perverse autocatalytic spiral: the topic becomes very popular and attracts more and more people in the pursuit of a rewarding subject in terms of citation return. Patently, this swarming of bees on the same beehive only ends up in stamping on each other's feet, in addressing the same problem over and over, and in duplicating things already achieved, perhaps claiming different and sometimes contradictory conclusions from known premises in an attempt to confer some novelty on one's work.

The final outcome results in such a mass of information that it becomes virtually impossible to disentangle the proper facts and to reach a shared point of view. All this creates an intense background noise, such as made by a conductorless orchestra tuning their instruments without ever beginning to play the symphony written on the score. Every now and then there are tuned notes and proper chords, but it takes a very well-trained ear to detect them out of the general cacophony.

To clarify the effects of citation indexes, I will tell a personal story. For some time, my studies have dealt with the electronic structure of the material class called oxides. During the 1990s my 'passion' had been a particular oxide that I mentioned earlier, magnesium oxide, MgO. On this topic I have produced some relevant results that over the years have accumulated many citations, reaching and exceeding 200 per paper. Then, at the beginning of the new century, I started to deal with another oxide, titanium dioxide, TiO_2, which has several practical applications and which, at the time, had not yet been fully characterized from the point of view of its electronic structure. Using the same techniques adopted years before for magnesium oxide, together with some of our experimental colleagues, we made some outstanding studies on titanium dioxide. Today, some of the resulting papers have collected 600–700 citations each, three times those collected on MgO. However, I can guarantee that those former papers are of the same scientific quality as the new ones. That is, it was enough to move from one compound to another to greatly improve our citation records.

Today, many people, the youngest in particular, choose their research theme also on the basis of such considerations, with the result that only a few devote their efforts to systems and processes with lower visibility, and greater risk. This again has a strict rationale. If no one is interested in a given topic, there must be a reason, and usually that is the case. But science must move in the domain of the unknown, and often the stubborn perseverance of those who for years have carried on with some unlikely idea forms the basis of sensational and unpredictable discoveries.

We can complain as long as we want about this state of affairs, but at the moment there is not much to be done. Around the count of papers, citations, impact factors, etc., the science of measuring science was born. There are new disciplines known as bibliometrics and scientometrics. Garfield, whom we introduced earlier, is considered their father, having founded the Institute for Scientific Information in 1960, and having created many other bibliometric products, such as Current Contents, the fundamental Science Citation Index, and other databases such as Journal Citation Reports. These were later complemented by similar databases such as Scopus, developed by Elsevier, or the powerful Google Scholar, entirely dedicated to research products. Despite the criticisms and the obvious limitations, already highlighted by Garfield in a critical article in 1979,[48] citations analysis has literally exploded in the past 20–30 years. Today, the field has its own scientific society, with researchers dealing only with this kind of problem.[49] It can be considered positive or negative, but it remains a tool that we will hardly be able to dispense with in the future.

Perhaps the most sensible opinion on famous impact factor has been expressed by C. Hoeffel:[50]

> Impact Factor is not a perfect tool to measure the quality of articles but there is nothing better and it has the advantage of already being in existence and is, therefore, a good technique for scientific evaluation. Experience has shown that in each speciality the best journals are those in which it is most difficult to have an article accepted, and these are the journals that have a high impact factor. These journals existed long before the impact factor was devised. The use of impact factor as a measure of quality is widespread because it fits well with the opinion we have in each field of the best journals in our specialty.

Tell me what your h-index is, I'll tell you who you are

For those who have never heard the expression 'h-index' before, it may sound a bit mysterious and vague, as if you wanted to label something that is not well understood, a phenomenon that escapes our control. It evokes expressions like 'X-rays', so called because of the mysterious nature of these radiations when they were discovered. Actually, the h-index is by no means mysterious and the letter 'h' has a trivial origin: it is the initial of the family name of the person who introduced it, Hirsch. In just over ten years of existence, this parameter has revolutionized the way people are ranked in scientific research. Even those who refuse to use it, perhaps rightly, ultimately take a look at it, just out of curiosity. Nowadays, no curriculum vitae of a person involved in research fails to include this number at the top of the list, along with their date of birth or their affiliation. No science database fails to show your h-index, next to the total of published works and the citations that these works have received. So what is this number that every researcher wants to see growing, along with academic reputation? Any blame (or recognition) should go to Jorge E. Hirsch, a physicist at the University of San Diego, California, who in 2005 published a paper destined to leave a mark.[51] In a study dense with mathematical equations looking more like a treatise on theoretical physics than on bibliometry, Hirsch proposed using a new bibliometric indicator. Suppose a scientist has published 100 papers throughout his career and that 30 of them have been cited at least 30 times each. This means that the remaining 70 papers have been cited less than 30 times, perhaps never. Well, this threshold line represents the author's h-index. Thirty of his or her studies are considered more relevant than the rest, on the basis of the fact that they are more cited (at least 30 times each); the others are less cited, and they can contribute to increasing our author's h-index only when some of them begin to reach and exceed 30 citations.

Hirsch's winning idea was to introduce a single number that reflects both the number of publications (quantity) and the number of citations per publication of a given author (quality). Like any brilliant idea, it is rather simple. It is clearly better than the crude enumeration of total published articles: 70 papers that no one ever cited produced no impact, probably. But 30 works that have given rise to at least 30 citations each have attracted some measurable interest from the scientific community. The success of the h-index relies on the fact that it is a simple indicator that provides imperfect, approximate, but synthetic information about the scientific progress of an individual. The index also avoids the loophole affecting the calculation of total citations: that is, an author who has published only a single highly cited paper with 10,000 citations while the rest of his or her publications are poorly or even never cited would have a total number of 10,000 citations because of that single, lucky publication, but an h-index of 1.

Obviously, there may be many reasons why a researcher has an h-index higher than another, without being smarter, more committed, or more deserving. In a word, without being better. One reason is that working in different sectors results in a different average number of citations: in medicine, as we mentioned earlier, the number is much higher than in other disciplines because of the higher number of scientists, and the h-index reflects this situation. There are also aspects related to age, as the h-index automatically rises with time and sometimes simply measures the seniority of a researcher (at the beginning of the career it is totally meaningless). Furthermore, many operators are strictly bound by confidential or industrial research where patenting counts more than publication. More examples could be given.

Despite all these shortcomings, shortly after being introduced the h-index was universally and generally adopted, becoming a mantra of contemporary science. While recognizing its limits, it is generally true that a high h-index reflects a scientist who is intensely active and whose work has some impact over time. A low h-index can have noble explanations and motivations, such as is the case for scientists who have decided to publish only a few works of very high level. However, these are more exceptions which, as often happens, prove the rule: for a senior scientist, a low h-index is often a manifestation of modest or even poor scientific production.

All this makes perfect sense, but there is a general problem with indexes, expressed some time ago by the British economist Charles Goodhart: 'when a measure becomes a target, it ceases to be a good measure.' This is known today as Goodhart's law.

Do not blame your fever on your thermometer

From all that has been said so far in this chapter, it is clear that the parameters that measure scientific productivity, such as citations, impact factor, and h-index, have become an integral part of the process of expanding scientific knowledge, with all the pros and cons that these things involve. Among the detractors of these instruments, some bring solid and acceptable arguments; others, the majority, often suffer badly from the introduction of indexes that mercilessly demonstrate their mediocrity and therefore would like to see them disappear *tout court*.

In doing this, they forget that if the patient has a fever, this is not to be blamed on the thermometer. The problem of contemporary science therefore is not bibliometry, but its uncritical and indiscriminate use. Scientists themselves often make good use of these tools, mainly due to a truth that is as simple as it is difficult to accept: the dimensions of the scientific enterprise have obtained, and perhaps surpassed, the threshold beyond which the process of producing knowledge becomes inefficient, unnecessarily expensive, and even socially harmful. In other words, we are many, maybe too many. The growing influence of bibliometric indexes is a fair picture of this truth, perhaps difficult to accept, but outstanding. It is a dramatic issue, very seldom recognized and discussed, which is what we will be doing in the next chapter.

6

Are we too many?

'I can't believe, not for a moment, that God plays dice!' Einstein shouted.

'Just stop', Bohr replied, 'telling God what to do with dice.'

The two gentlemen had been discussing animatedly for two days. For a few years, attention had been focused on a new physics, still obscure and far from our perceptions. It had been formulated by such people as Max Planck, who at the beginning of the twentieth century had introduced the concept of quantized amounts of energy, the photons, which was then translated into the modern concept of the atom by the Danish physicist Niels Bohr, the father of modern atomic theory. But then others joined in to reinforce this new theory, indispensable to explaining the phenomena of the atomic world: Louis De Broglie, with his intuition that matter can have both particle and wave nature; Erwin Schrödinger, whose wavefunctions were devised precisely to describe that evanescent world; not to mention a noncommittal and bizarre character like Paul Dirac and his complex formulation of quantum mechanics. But nothing had disturbed Einstein more than a principle put forward a few months earlier by Werner Heisenberg, according to which it is impossible to measure at the same time the exact position and the precise speed of a particle: there must always be a small residual error, an indeterminacy. Briefly, one could no longer speak of certainties, but only of probabilities. Bohr had completely endorsed this revolutionary vision, but Einstein, who as far as

The Overproduction of Truth. Gianfranco Pacchioni.

DOI: 10.1093/oso/9780198799887.001.0001

revolutionary visionaries are concerned is certainly unsurpassed, was not convinced at all.

A proper occasion to discuss such matters came in October 1927 when Einstein and Bohr met in Brussels for the fifth Solvay Congress, devoted to 'Electrons and Photons'. Every morning at breakfast Einstein presented Bohr with an imaginary experiment designed to tear to pieces Heisenberg's uncertainty principle. Bohr would analyse these challenging hypotheses until he was able to find a counter-argument to retort. Nonetheless, Einstein would come back the next day, quite at ease, with a new, different imaginary experiment. One morning Einstein came up with an idea that he considered winning. He figured out a box from which a ray of light comes out at a precise moment. By weighing the box before and after emitting the beam, and by exploiting the famous $E = mc^2$ equation that connects mass and energy, Einstein concluded that it was possible to derive the energy of the emitted radiation, and that by knowing the exact moment when the ray left the box, it was possible to violate the uncertainty principle!

It was a pretty convincing construction, and Bohr was deeply concerned: this thought experiment was shaking the foundations of the atomic model of quantum mechanics and he could not get along with it. Very upset, he tried all day long to convince the other participants at the conference that things could not be as Einstein had set them out. One of the attending scientists so described the scene: 'I will never forget the picture of these two top scientists as they left the club: Einstein, tall and authoritative, walking quietly with an ironic and contented smile, and Bohr staggering after him, very nervous.' Bohr spent the whole night trying to disassemble Einstein's experiment and the next morning he came down to breakfast, triumphant. He had found an escape based on nothing less than the theory of relativity, the very theory that Einstein had formulated a few years before. Bohr proved that the gravitational force required to weigh the box would have

Figure 4 Group photo of the 5th edition of the Solvay Conference on Electrons and Photons, held in Brussels in 1927 at the Institut International de Physique Solvay. From left to right and from top to bottom: A. Piccard, E. Henriot, P. Ehrenfest, E. Herzen, Th. De Donder, E. Schrödinger, J. E. Verschaffelt, W. Pauli, W. Heisenberg, R. H. Fowler, L. Brillouin; P. Debye, M. Knudsen, W. L. Bragg, H. A. Kramers, P. A. M. Dirac, A. H. Compton, L. de Broglie, M. Born, N. Bohr; I. Langmuir, M. Planck, M. Sklodowska-Curie, H. A. Lorentz, A. Einstein, P. Langevin, Ch. Guye, C. T. R. Wilson, O. W. Richardson.

influenced the flow of time, as predicted by Einstein's relativity, blurring the measurement of the precise moment at which the light beam would leave the box. Einstein's conjecture was refuted! And it was at that point that Einstein lost his patience: 'God does not play dice', he blurted out, unwilling to surrender to a view in which the description of nature could only rely on probabilistic laws, without absolute certainty. The debate went on for years; Einstein never became fully convinced, but neither was he able to find a way to dismiss the uncertainty principle, which was shortly to enter with flying colours into the fabric of modern quantum theory.

If I report this story, it's because it introduces an impressive overview of how the community of scientists has changed over time. Look at the picture in Figure 4, taken at the famous Fifth Solvay Conference (24–7 October 1927) where the above dispute took place. There are 29 participants. Of these, 17 had been awarded the Nobel Prize at the time of the conference or were to be awarded in the years to come. These include not only Einstein, Schrödinger, Heisenberg, Dirac, de Broglie, Bohr, and Planck, mentioned above, but also Madame Skłodowska-Curie, Pauli, Debye, Bragg, and others—people whose discoveries are now in the textbooks of physics students all over the world. They are all there, in beautiful poise. Something like the Mount Olympus of physics of the first decades of the twentieth century, people who contributed to the 'thirty years that shocked physics', as goes George Gamow's famous expression.[52] All at the same conference, trading their wisdoms in the challenge of reaping nature's secrets about the quantum world.

Midget versus giant congresses

Today, a regular congress of the American Physical Society, just to stay in the field of physics, sees the participation of about 10,000 people. Four such conventions are held every year, each devoted to a different theme. The two annual meetings of the American Chemical Society attract between 13,000 and 18,000 participants each. Still few compared with the Congress of the Society for Neurosciences, approaching some 30,000 attendees.[53] It is a case of meetings whose size is 300 to 1,000 times that of the 1927 Solvay Congress. Some Nobel Prize winners are still present, but 'diluted' in a sea of 'normal' participants. If you think these numbers are out of this world, you'd better change your mind. The largest scientific congresses are held in the field of medicine; just to name one, that of the Radiological Society of North America, in 2016, saw the participation of 52,000 people.[54]

In these mega-congresses most of the scientific communication takes place in the form of posters, that is, a sort of *dazibao*, the old, traditional handwritten means of protest and popular communication in China. In these printed panels, in less than two square meters, are condensed the results of months and sometimes years of work. Authors stand beside their posters for hours during the special poster sessions in the hope that someone from the wandering crowd will be attracted by a figure, by a title, by the author's name, stopping by to have a chat and be told in more or less detail about the achieved results. It is a kind of open market where information is exchanged, and business is done, among graduate students looking for postdoctoral positions, post-docs considering their next job experience, team leaders looking for good candidates for open positions, and so on. The goods are not for sale, being just on show. Information is exchanged on how to run a given experiment or handle an instrument, how to perform a certain measurement, how to synthesize a specific compound. For the few, luckiest ones there are oral communications, 15 minutes long, sometimes 12 or even less, fleeting times in which one must squeeze in introduction, motivation, state of the art, methods used, results, conclusions, and acknowledgments to supporting agencies: a real exercise of conciseness and brevity. Only those who have a wider reputation or have at hand some 'hot' results have the honour of giving a half-hour invited talk. And above all stand the few, exceptional, almost sanctified 'big names', those who speak in plenary sessions in front of thousands of more or less careful listeners who dream of one day being able to swap places with the speaker at the lectern.

Between the image of the 1927 Solvay Congress, small but rich in talent, and that of the mega-congresses of present day, crowded *kermesses* somewhat reminiscent of a Middle Eastern *souk*, is condensed the whole transformation that modern science has undergone in recent decades. The picture is clear. Science has grown immensely over the years, so much so that it has changed from the trade of a few highly motivated and passionate adepts into a flourishing global industry with millions of people engaged in scientific work, managing to produce 25 million scientific papers in the time span between 1996 and 2011, according to a 2014 survey.[55]

And so we come to grips with the central theme of this book, the embarrassing question I asked in the Introduction, maybe a provocative consideration, but one that, I am sure, is well worth tackling: are we too many? I'm talking about the global community of scientific people active in research, white, yellow, black, and red scientists. Science has always been universal, without frontiers and transnational. More importantly, the problem of the growth of researchers in China and India cannot be disregarded by those working in Germany and the United States. But how many active scientists are there exactly?

How many are we?

The answer, seemingly simple, is actually not so easy. In fact, there seems to be no unique certified and accepted value, mainly because of the difficulty in defining exactly the profile of a researcher and scientist outside the academic world (the qualification of R&D scientist in the industrial world is much more evanescent). The British Royal Society reported an estimate based on UNESCO data, according to which there were 7.1 million scientists in 2007, up by nearly one and a half million compared to 2002 when their number was estimated at around 5.7 million.[56] In 2011, the publishing house Elsevier conducted a survey commissioned by the British Government, attaining a value of 5.95 million active scientists in 2009.[57] These data refer to the definition of researcher cited in the *Frascati Manual*, a document that assesses the methodology for collecting and using research and development data in OECD member countries:

> Research and experimental development (R&D) comprise creative work undertaken on a systematic basis in order to increase the stock of knowledge, including knowledge of man, culture and society, and the use of this stock of knowledge to devise new applications.[58]

This is more accurate and analytical than the definitions of scientists and engineers used by UNESCO: for this reason, for example, the total number of people working in the field of scientific research in China in 2006 ranged from 1.1 to 1.6 million, depending on whether you used the more restrictive definition of the *Frascati Manual* or UNESCO's more general definition. According to another survey, this time by the National Science Foundation, the number of researchers rose from about 4 million in 1995 to about 6 million in 2008.[59] Of these, about 25 per cent work in the USA (1.4 million) and 25 per cent in the European Union (1.5 million, 44 per cent of which in the private sector, 12 per cent in government agencies, and 42 per cent in academia).[60] The scenario is evolving. Growth rates differ widely from country to country, and bring about major variation in the composition of the worldwide floor of researchers. The United States had a yearly growth rate of 3–4 per cent between 1995 and 2002, after which growth continued at lower rates, around 1 per cent per year. Countries like Russia and Japan have seen a levelled number of scientists in recent years, while India went from 357,000 researchers in 1995 to 441,000 in 2010. China, on the other hand, grew by 6–7 per cent yearly between 1995 and 2002, and in subsequent years the numbers blew up by astonishing two-digit factors, 10–12 per cent a year. And so, according to UNESCO data, China went from 804,000 researchers in 1995 to 3,250,000 in 2013, a fourfold increase in less than twenty years. Fast growth is not only China's prerogative; in South Korea the growth rates are only slightly lower (see endnote 59).

An immediate consequence is that China has already exceeded the combined number of researchers active in the United States and Europe (see endnote 27). It is no surprise, then, that nowadays the number of scientific papers in English language coming from China is larger than that coming from the traditional cradles of scientific research, Europe and the United States.[61]

We come now to a final estimate of the total number of currently active researchers. According to the latest available data from the National Academic Press,[62] in 2013 there were around 9,500,000 researchers in the world, not counting the USA. Adding to this figure the latest estimate of active scientists in the USA, i.e. 1,400,000, we have the total of 10,900,000. On the other hand, by extrapolating the complete UNESCO data of 2002 and 2007, assuming continual growth at a steady rate of 4.5 per cent per year, we estimate there were 10.4 million in 2016. The two values are quite similar. We certainly have exceeded 10 million people involved, in one way or another, in scientific research. And we reckon that about 90 per cent of scientists of all times are alive and active today. The number of authors differs, however, primarily because by no means will all of these publish an article in a given year. For example, 2.4 million articles were published in 2013 by a total of 4.16 million unique authors (see endnote 7).

These are the numbers. But how many scientists were active 30, 50, or 100 years ago? Clear and trustworthy data do not seem to exist, but we can try to make some rough guesses. Assuming a constant 4.5 per cent yearly growth over the past few decades, and starting with about 10,000,000 bona fide active researchers in 2015 (broadly speaking, not just academics), it can be deduced that there were about 4,000,000 scientists in 1995, 1,700,000 in 1975, 700,000 in 1955, and less than 300,000 in 1935. We can then cross-check these estimates with some data reported in a famous article by Max Perutz, a British molecular biologist, Nobel Prize winner in 1962 for his studies on haemoglobin and myoglobin. In an

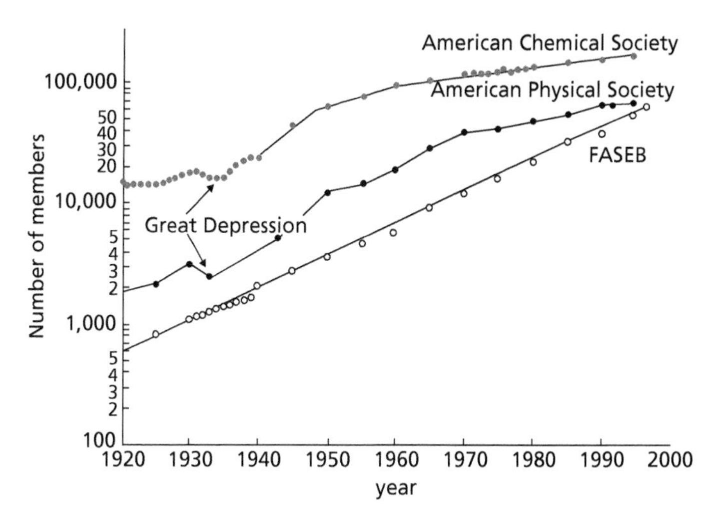

Figure 5 Growth of members of some American science societies in the past century (note the semi-logarithmic scale). FASEB stands for Federation of American Societies of Experimental Biology.

Source: Max F. Perutz, Will biomedicine outgrow support?, *Nature* **399** (1990), 299–301.

article entitled 'Will Biomedicine Outgrow Support?', published in *Nature* in 1999, Perutz commented adversely on the fast and unchecked growth of biomedical researchers.[63] To support his contention, Perutz showed, in a graph (Figure 5), the evolution of the number of members of three major American scientific societies, those of chemistry, physics, and the Federation of American Societies of Experimental Biology.

The data are interesting. In 1930, the members of the American Chemical Society were about 18,000; physicists were much less numerous, for a total of around 2,000. Even less were the associate biologists, who barely reached 1,000 members. In total, these three categories amounted to more or less 20,000 registered members of their respective scientific societies. In 1965, or 35 years later, chemists had become about 100,000, physicists 30,000, and biologists 10,000, which means that numbers had increased almost sevenfold globally. Growth continued, albeit with a lower rate in chemistry and physics, about 2 per cent from the 1960s to the end of the century, while the bio world was proliferating at a constant rate of about 5 per cent per year. By the end of the century the members numbered about 150,000 chemists, 70,000 physicists, and as many biologists, for a total of nearly 300,000 people in charge of research. Obviously, these numbers are very partial, covering only members of American scientific societies and only in these few domains. On the other hand, in the past century, it can be said that scientific research was mainly a prerogative of the United States and of European countries (not all of them, though: for example, Spain rose fast only between 1980 and 2000).

So, in 1930 there were something like 20,000 members of the major American science societies. We can reasonably estimate that there were no more than 200,000 scientists in the whole world, probably much less than that. At that time the world's population was 2 billion people, which means that there was one researcher every 10,000 people. In 1960 scientists had become many more, but they had not reached one million. The world's population had risen to 3 billion, and the ratio had become one researcher every 3,000 people. By the end of the past century the number of active researchers had reached and exceeded 5 million: in the meantime the world's population was 6 billion, thus coming close to one researcher every 1,200 people. In 2012 we earthlings became 7 billion, with just under 10 million scientists: roughly, one every 700 people. In 2048 the projected world population will be 9 billion, and if the number of scientists keeps growing at the annual rate of 4 per cent, at that date they will be 35 million: one in 250! This was already forecast in 1994 by David Goldstein: 'It is a mathematical fact that if scientists continue to multiply more rapidly than the population, there will soon be a time when there will be more scientists than people.'[64] Goldstein, however, concluded that the growth curve was already bending downwards—which then turned out to be wrong (demonstrating that scientists too may be wrong) because he did not take into account the 'explosion' of researchers in Asian countries that was just beginning in those years.

The steady and tumultuous growth in the number of researchers is the consequence of an equally strong increase in global R&D investment. According to a recent estimate (see endnote 6), in 2013 the allowance had reached almost $1.7 trillion, doubling the $836 billion allowed ten years earlier (Figure 6). This expense is essentially concentrated in three geographical areas: East and Southeast Asia, North America, and Europe. The USA remains the largest R&D investor with 27 per cent of the world's expense, followed by China with 20 per cent. Also the trend is interesting: between 2000 and 2013 the expenditure in Asian countries rose from 25 per cent to nearly 40 per cent of the world's total, while in North America and Europe it has dropped.

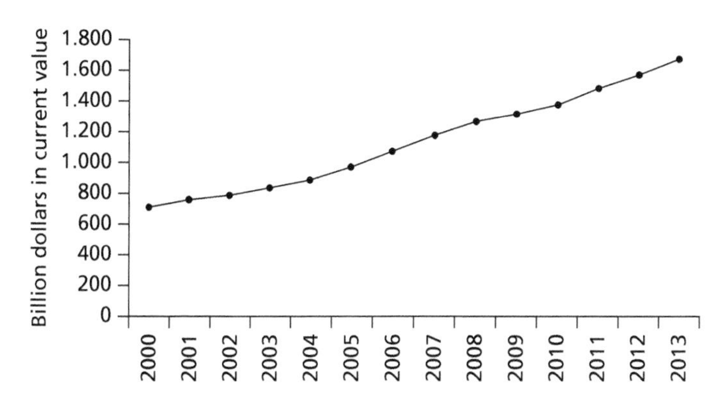

Figure 6 Overall research and development spending, 2000–2013.

Source: Science & Engineering Indicators 2016, Washington, DC, National Science Board, National Science Foundation.

The question becomes: do we really need all those scientists? It seems obvious that if the population doubles, there is also the need for double the number of orthopaedists, barbers, and city policemen. If your pipes burst in Kuala Lumpur, they may hardly be fixed by a plumber from Cardiff. These services are independent from one another, and demand growth along with population. Living as we are in an increasingly technological world, it is not surprising that there are more and more people with technical and scientific qualifications, involved in the implementation, development, and optimization of these technologies. They are all scientists and engineers, and their number is going to increase with the spread and growth of high-tech companies and activities. But in fundamental scientific research this is not the case.

Science is universal by definition. If a Seoul scientist finds the answer to a problem, there is no longer a need for a scientist in Rio de Janeiro to solve the same problem. At most, the scientist in Rio de Janeiro can confirm the result obtained in Seoul. What is more, if it is true that the most significant progress in the past came from 'blue sky', free and unconstrained research, one must also admit that this progress must be credited to dedicated, talented people, who had been struggling to solve very complex problems. Today, the danger is fragmentation, that is, consideration of more and more minute, less and less relevant issues. In conclusion: perhaps we do not need as many 'pure' scientists as plumbers. But—once again: clearly, this reasoning holds for fundamental research, free in concept, and free from direct application purposes. In a world where technology has a constantly growing role, there is an obvious need for an expanding array of technicians, specialists, and well-trained people capable not only of developing new processes and new technologies, but also of applying, managing, and improving them. But this is apart from fundamental research, where only talent, genius, and originality, not so much workforce, should make the difference. True, today large scientific projects (and even smaller ones) can require a good number of skilled hands to produce relevant results in reasonable times. Also fundamental research requires workmanship that, however, should be employed to tackle only really important, original problems and challenges.

Are we churning out too many PhDs?

A few years ago I was invited to give a talk at a conference on the electronic structure of solids and materials in Berlin. Going back to the city where I had lived for four years in my youth always makes me happy, and it was in high spirits that I approached the Henry Ford Bau, the large white building in the Dahlem district where the congress was being held. Meanwhile, Berlin had become a very fashionable city and, in fact, the congress had attracted a very large number of participants in our field: 1,200. After giving my talk, in the afternoon I decided to spend some time in the poster session, although these are always very noisy, crowded with plenty of people and young students who want to tell you every little detail of their work when you would prefer to sit quietly in front of a beer. But my lazy side succumbed to the active one, and I plunged into the crowd. It was really an impressive poster session. A long sequence of panels all of the same size, each with a beautiful poster with nice colour images, had been unbelievably crammed in two full floors of that huge building.

In front of each poster stood the authors, sometimes just one, sometimes a few young scientists. The unifying theme of the conference was the calculation with advanced methods of the properties of new materials. I walked around, diving deep into that loud crowd, looking from a distance and rather superficially at one poster after another, when I suddenly realized that many of these studies had a high scientific content, a wealth of results, a depth of analysis, which had nothing that would be envious of what I had shown in my morning oral presentation. In other words, I realized that many of these young people could have given an invited talk in my place, maybe reporting results even more interesting and more innovative than mine. Only age and reputation had given me an edge. As I was proceeding through the dense crowd, I became increasingly impressed by the number of young talents I was meeting: all motivated, competent, all with already important results in their CVs, but all more or less in the same condition: namely, hunting for a minimum of visibility, for a chance to emerge, or even for a chance to transform their numerous and intense efforts into a permanent research position at a university or research centre.

At that moment I realized that for many of them, indeed for most of them, these chances were close to zero. In front of me were a host of brilliant young scientists, all involved in the same subject, all with the same expertise, almost all Europeans. Obviously, they were only a part, a small part of those who dwell in that specific research area in the rest of Europe and in the rest of the world. Even with the best intention of looking at things with confidence and optimism, it was impossible to imagine for all those bright young men and women a shining future in the academic world. When I arrived at the 30th poster glaring with fantastic results, I decided that it was too much, and with a pinch of sadness, I proceeded towards my highly desired and well-deserved beer, but with a clear feeling that something was moving the wrong way. Perhaps my generation was pushing too many young people into a world that would never be able to welcome them all. Perhaps we were producing too many PhDs, or at least too many would-be full-time scientists in the academic world.

For many years it has been taken for granted that scientific and technological progress can only be achieved by producing more scientists. True, I used to think so too, and in a sense I still think so. At the same time, it so happened that those who decided to engage in that activity had a reasonable chance of being able to keep practicing for a long time and to become professional scientists. Obviously, certainties never existed, the path has always been rough and bumpy, but at least there were some opportunities. Twenty years ago, it was commonplace in the USA to be offered a tenure-track position (a position soon to become permanent) at some universities just a couple of years after earning a PhD. Today, the fate of many, if not all, of those who after their PhD want to undertake an academic career is to set out in an uncertain hunt for temporary positions, seldom lasting more than 2–3 years: the usual post-doctoral positions, postdocs in the jargon of the trade. A first postdoc is usually followed by a second, and then by a third, continual jumps in uncharted land, often in different cities, in different countries, in different continents. Years go by, self-assurance vanishes, the risk of never finding a permanent placement increases. It's not really an idyllic picture.

From endemic to pandemic

Today, the structure of academic research is more or less the same all the world round There is a group leader, usually a professor assisted by some experienced researchers, with a (mostly, but not always) permanent position coordinating a number of postdocs of variable seniority, each of whom in turn is responsible for one or more PhD students. It is not uncommon to find groups where, depending on the amount of funding, there may be 10 or even 20 postdocs at the same time. It is a real pyramid, which stands until the number of jobs grows so as to absorb the people that were trained in such a process.

Someone has tried to estimate the effects of this system using the growth factor R_0. In demography R_0 is defined as the average number of girls born to a woman during her life. A factor larger than 1 indicates that the population will grow over time, and vice versa. In epidemiological studies, R_0 is the average number of people infected by a carrier during the incubation period. A disease with an R_0 factor greater than 1 generates an exponential growth of infected people and an epidemic occurs. Simply said, a normal flu has an R_0 value of around 1.2, while an R_0 value of 4 has been estimated for the devastating Spanish influenza of 1918. In a publication in *Systems Research and Behavioural Science* in 2014, Richard C. Larson and colleagues introduced the R_0 factor as an index of growth in the number of PhD students.[65] R_0 was defined as the average number of PhDs that a professor delivers during his or her career. So, R_0 equal to 1 means that a professor will produce only one PhD in his career that will have access to a permanent place in the academic world, thus maintaining the population in a stationary state. Analysing data in the field of engineering, the authors show that in the USA $R_0 = 7.8$, which implies that, under stationary conditions, only one doctoral student out of 7.8, i.e. 12.8 per cent, are likely to get an academic position. In prestige institutions such as the famous Massachusetts Institute of Technology (MIT), R_0 approaches 10. Clearly, if all doctorate students were looking for an academic career, the system would collapse.

Fortunately, most PhD students find employment in industry, hospitals, financial markets, and, in general, in productive society. This is also the main goal whenever a PhD programme is undertaken: acquiring research and innovative skills to be transferred to companies, to the economic system, and to society in general. But basic research is a very attractive activity; those who work in science often practice it with dedication and enthusiasm, and all this is transmitted, without filters, to doctoral students who are driven, directly or indirectly, consciously or unconsciously, toward the replication of the academic career of their mentors, at least in intention.

Very seldom during their PhD years are students warned about the dangers and uncertainties of attempting to engage in academic research, so that the number of postdocs trying to beome scientists grows faster than that of prospective PhDs. A study from the National Science Foundation describes the situation well. In the USA in 2015 there were 685,000 doctoral students in engineering, science, and medicine, almost twice as many as in 1975 when they were 328,000.[66] In the same period and in the same disciplines the number of postdocs had grown from 18,101 in 1975 to 63,861 in 2015. Thus, in forty years while the PhDs had doubled, postdocs had grown by a factor of 3.5. The question of when and why to look for a post-doctoral position, and with what risks, has begun to spread around.[67]

It is estimated that in the United States, only 25 per cent of PhDs in science end up obtaining an academic position, and even less, about 15 per cent, manage to get a permanent position.[68] Things are even worse in the UK where it seems that no more than 4 per cent of PhDs stabilize in the academic world. There is the added consequence that the average age at which a researcher obtains a stable position in the USA is about 42, an age at which in the past many scientists had already been awarded a Nobel Prize. Competition for the few openings has become so strong that for a single position, hundreds of applications are received, making the selection process very complex and the success rate desperately low.

Under pressure

What has been outlined so far has resulted in a lot of pressure on those young people with a strong desire to become part of the world of research. Foremost are PhD students, who must emerge into a not only very crowded, but extremely competitive environment. Given the high number of postgraduates and postdocs and the scarce number of available academic positions (at least in the

Western world; China, with its strong growth, is a different story) only a narrow minority has a chance of making any progress. In this competition, time is crucial. Scholarships, academic awards, and career advancement all depend on two essential factors: publish well and publish quickly. Publish well means placing your work in high impact factor journals. But to this end it is also indispensable to produce results far above average, and the time factor as well as a bit of luck play a decisive role. A student who happens to be in the right group at the right time when a discovery is made can enjoy all the benefits. But the same student, with the same qualities, can just as well get trapped in a group at the wrong time, when problems of internal balance or contingent instrumental difficulties arise, and this may have a dramatic influence on his or her scientific production during the PhD. The three to five years of activity needed to acquire the doctoral title become therefore decisive: you may be lured by the illusion of a brilliant career or you may see your expectations dramatically collapse simply due to external or random factors.

The time at hand to demonstrate your qualities is short and there is no room for errors or second thoughts. Young people feel pushed to perform miracles to have their results published in prestigious journals. But publishing in high-impact journals is not only difficult, it also takes time, often months of disputes with referees and editors, and the whole enterprise can eventually end up in a rejection, which results in frustration, stress, and fear of not being up to the challenge. Pressure to achieve exceptional results can also come from the broader environment: sometimes from the group leader, who in turn must consolidate his position by moving from temporary to permanent staff; sometimes from postdocs themselves, in desperate search for the big hit that will grant them a bright future, but see their chances getting slimmer and slimmer as time goes by. In such a predicament, interpreting one's results in a 'benevolent' way, finding some exciting result at any cost with the hope of pleasing the boss or of increasing your chances for the future, can be an irresistible temptation. Controls become loose, the hurry to publish leaves no space to repeat and double-check the measurements, any doubtful result is simply discarded, deleted, forgotten. Cutting corners, doctoring data, plagiarism, and even plain fraud find in this humus their most fertile ground.

It, thus, becomes more likely to take a narrow, steep, descending path, from which it becomes increasingly difficult to go back, with the danger of leading down to the final disaster. It's the story we're about to tell. It is the story of how pressure to emerge can lead to total destruction and total loss. It is the story of the scientific fraud of the century.

7

Famous frauds

On the morning of 3 May 2002, Liesbeth Venema, senior editor of *Nature*, went as every day to her office in London near King's Cross station. It was a day like any other in London, gray and looking like rain, although some patches of clear sky were not to be ruled out for later on. Waiting for Venema on her office computer were emails that had arrived during the night, especially those sent from the United States, due to the time lag. It was the usual mixed batch of referee responses, new paper submissions, lists of changes suggested by referees, inquiries about the status of submitted papers, and so on. But the email sent by Lydia Sohn, a professor of physics at Princeton University, was definitely something different. The text was synthetic, and was commenting on a slide in a PowerPoint file attached to the email. In that slide Sohn had spliced together two figures from two articles by the same author that had appeared a few months earlier, one in *Nature* and the other in *Science*, that is to say, the two most important journals in the world's scientific literature. As Sohn pointed out, a most disturbing thing was that the curves shown in the two figures, referring to different experiments and published in two competing journals, were identical!

As she struggled to recover from the shock, Venema began to ponder what to do. She did not yet know that she was witnessing the start of an avalanche, beginning to rumble down, that would eventually overturn an outstanding piece of top physics and nanotechnology research of the past three years. It was the

The Overproduction of Truth. Gianfranco Pacchioni.

DOI: 10.1093/oso/9780198799887.001.0001

beginning of the end of a story that had captured the attention of larger and larger sectors of the scientific community, a series of amazing discoveries that seemed to open the way to new, extraordinary technologies. That email marked the 'game over' point, a game that had been played by a young and brilliant scientist who in the previous months and years had piled up an impressive set of sensational discoveries. Venema had in front of her for the first time positive proof that something was wrong. This was not yet a DNA match of the culprit, but something very close to it: she was handling the first explicit disclosure of one of the biggest and, since then, most discussed scientific frauds ever.[69]

I have chosen this particular case to analyse the topic of fraud in science for three reasons. The first is that this happened in the fields of microelectronics and nanotechnology, very close to my research interests. The second is that I became aware of that 'crisis', albeit indirectly, through personal contacts while it was taking shape. The third is that by size, depth, and impact, the case is perhaps the most extensive scientific fraud ever encountered, and analysing how it could have happened helps to understand the mechanisms by which modern science proceeds. There is no need to reaffirm that cheating in science has always existed, but it must also be said that it is very rare, although, unfortunately, on its way to becoming more frequent. In recent years, several surveys reporting well-documented cases of fraud have been published.[70] But although they stir lots of media hype, such cases stay quite few in number for a very simple reason: science has control and verification mechanisms that sooner or later disclose whether a result is genuine, manipulated, or plainly false. Generally, this happens before extensive damage is done with, God forbid, a malfunctioning device or a fake drug being put into production and hitting the market. In this respect, science is quite different from other, ordinary human activities in which the discovery of a fraud often comes too late, when things are well on their way to dramatic consequences.

In 2003, the world was told of the absolute necessity of waging war on Saddam Hussein's Iraq, to dispose of lethal weapons of mass destruction that the dictator had purportedly accumulated in his secret arsenals. Only much later was it discovered that these weapons had never existed, and that the world had been somehow cheated. But the damage had been done: war broke out and a whole part of the world was destabilized for years to come. In science such a thing is unlikely: if an important result has been falsified, sooner or later this will come to light. Reproducibility of results has been the foundation of modern Galilean science—at least until now. If a result is significant and of potential impact, then many groups will embark on the attempt to reproduce the detailed paths or procedures and, if possible, to improve them. If the claimed result is a fake, there is no escape: the imbroglio will come to light. For this reason, scientific cheating is a rare bird and, from a certain point of view, it defies understanding. In fact, nothing in such cases is it more certain that crime does not pay. There is one major exception: when a manipulated or falsified outcome is irrelevant and therefore bears little impact on the community. In this case, why waste time and money checking or duplicating a result of no interest? But that is another issue having to do with the mass of scientific information of little or no interest—'not even wrong', as the saying goes—which permeates the world of scientific communication.

But let's go back to the incredible story of Jan Hendrik Schön, the young German physicist who plays the role of protagonist in this drama.

From PhD to Bell Labs

Hendrik Schön conducted his doctoral studies at the University of Konstanz, a small, pleasant town on the lake of same name, straddling the border between Switzerland and Germany. There he graduated in 1997 working on the properties of organic crystals.

In those years one of the hot topics in physics research was the possibility of using organic substances, such as sugar or naphthalene, to produce new electronic materials. Computers, modern information technology, the Internet, and satellite telecommunications are all based on an inorganic semiconductor, crystalline silicon, of strategic importance for the development of transistors and integrated circuits. The search for smaller and smaller systems, aiming at lighter and more flexible electronics, required the development of new organic-based materials. Schön's research was broadly involved in this research line, and his doctorate produced some interesting results. Thanks to the contacts that his PhD supervisor, Ernst Blucher, had with a famous Bell Laboratories scientist, Bertram Batlogg, at the end of 1997 Schön had an opportunity to migrate to the United States to work in one of the most prestigious research centres in the world.

No less than eight Nobel Prizes in Physics and Chemistry enlighten the long, successful history of the Bell Labs. The first came in 1937, when Clinton J. Davisson shared the prize with O. Germer for the demonstration of the wave nature of matter. Perhaps the most important was the second in 1956, when John Bardeen, Walter Brattain, and William Shockley received the prize for the invention of the transistor, one of the milestones in the history of Bell Labs but also of contemporary science. Twenty years later, Philip W. Anderson was awarded the prize for his research on the structure of magnetic materials. Just one year later the quality of research at Bell Labs was newly venerated with the 1977 prize being awarded to Arno A. Penzias and Robert W. Wilson for the incredible discovery of cosmic background radiation. A jump of twenty years to 1997, the year in which Schön graduated and then arrived at Bell Labs, Steven Chu received the Nobel Prize in Physics for developing a technique that cools and traps atoms through a laser beam. The following year, Horst Störmer, Robert Laughlin, and Daniel Tsui were awarded the prize for discovering and explaining the fractional quantum Hall effect. In short, Schön had ended up in one of the largest physics centres in the world, a true science temple where 'historic' Nobel Prize winners dwell alongside fresh awardees, together with a strong group of potential laureates. The most recent awards have been the Nobel Prize in Physics in 2009 to Willard S. Boyle and George E. Smith for the invention of an image sensor, and the 2014 Prize, this time in Chemistry, to Eric Betzig for his work started at Bell Labs on fluorescence microscopy.

Surely Bell Labs represented a very fertile ground for promoting the qualities and talents of a young scientist, being, however, also a place where competition was pushed to the extreme. The task assigned to Schön was to replace traditional semiconductors based on silicon with organic materials. In the form of polymers, organic compounds are best known as plastics, which led to the ambitious goal of producing a revolutionary 'plastic electronics'. The problem is that the electrical conductivity of organic substances is hundreds or thousands of times lower than that of silicon. The aim of the research was to develop crystals of organic substances whose ordered structure would sustain an unprecedented electronic mobility, so as to produce a real organic conductor. The first results obtained by Schön concerned the construction of a field-effect transistor where one of the components was made of an organic material. The results were reported in a Communication sent to the *Journal of Applied Physics*. However, the paper was rejected by the reviewers because the measurements reported in the article had raised more than an eyebrow. In September 1998 Schön resubmitted the work to the same journal, including new data intended to dismantle the referees' objections.

With these small changes the paper was accepted for publication. Later, however, it turned out that at least one of the figures had been artificially modified by adding data by computer editing, without actual measurements. In some ways Schön, prompted by the requests of the referees, adjusted his results to counter their objections. After all, it was very easy: adding a few points on a graph was enough to turn unconvincing work into a publication of interest—and, surprisingly, nobody noticed the trick. With such premises, probably, Schön began to walk down the slippery path that would lead him to perdition.

'Plastic fantastic'

At the beginning of 1999 Schön was working on a manuscript introducing the possibility of creating photovoltaic solar cells based on organic materials rather than on the traditional silicon. The study, submitted to *Nature*, sailed in hostile waters but in the end it was accepted. It was Schön's first paper published in such an important journal, and for Schön that was an extraordinary achievement (but that work, too, contained some manipulated data, as was discovered after the scandal broke).

Schön's productivity in those months was really impressive. Having kept in touch with the group where he had defended his doctoral dissertation at Konstanz, Schön regularly went to his former laboratory to operate an instrument designed to deposit thin layers of oxides, especially aluminum oxide. According to Schön, no one else was using that device, and this was offering a splendid opportunity for him to develop new materials for his work at Bell Labs. In particular, with that machine he experimented with depositing a thin layer of aluminum oxide on the surface of pentacene crystals, an organic compound. That should have been a crucial stage for the development of a new generation of transistors. In fact, what makes it possible to use silicon as a semiconductor in a transistor is the fact that a thin layer of insulating material, in this case silicon dioxide, is deposited on its surface. Accordingly, also organic crystals, in order to function in a transistor device, must be covered with a thin layer of an insulating oxide. Unfortunately, this operation is extremely difficult with organic crystals because their structural and mechanical properties are very different from those of silicon. Schön began to experiment with this complex problem, which, as we shall see, will play a pivotal role in the whole affair.

Meanwhile, another very important achievement was in store. At the end of 1999, working with tetracene crystals (quite similar to pentacene), Schön measured a phenomenon known as the quantum Hall effect. The effect occurred at a very low temperature, 1.7 degrees Kelvin, just above the temperature of absolute zero. The same effect, observed in 1980 in silicon, was worth the 1985 Nobel Prize in Physics for its discoverer, Klaus von Klitzing. Schön chose to show his results to Batlogg, his boss at Bell Labs, on the morning of 23 December 1999. The choice was not casual. It was, in fact, a historic occurrence. As is well known to those working in the field, on the day before Christmas Eve in 1947 at Bell Labs Shockley, Bardeen and Brattain showed the director of their lab the first example of a functioning transistor. It was the beginning of the electronic revolution, and that date is a hallmark in the history of science and technology. Fifty-two years later, on the very same day, Schön announced to his bosses a major discovery that made organic crystals more and more similar in performance to mythical silicon.

A boy with golden hands

It really seemed that Schön, the promising boy, had found a sort of new philosophers' stone: he did not change metals into gold, no; however, he had made his ultra-pure organic crystals of golden interest in a growing set of applications in condensed matter physics. In the spring of 2000, Batlogg left the Bell Labs to take over the leadership of an important laboratory at ETH, the Zürich Polytechnical School. This allowed Schön to expand himself in terms of space and autonomy. His colleagues saw Schön almost always at work in his office, intent on typing on his computer. Someone started wondering about when Schön would be doing his wonderful experiments, but it is not unusual to find researchers who prefer to work in the strangest moments, in the middle of the night as well as on weekends, when it is quiet and fewer people are around. The fact that Schön, an experimental physicist, was spending so much time in his office sitting in front of his computer did not seem exceedingly strange. His production then continued at a fast pace, such that in the course of 2000 only, Schön submitted five papers to *Science* and three to *Nature*, all as corresponding (i.e. main responsible) author.

It is no surprise, then, that the reputation of the young German researcher began to spread rapidly. In April 2000 Schön was invited to Stuttgart by the Nobel Prize winner Klaus von Klizing, the discoverer of the quantum Hall effect. In Stuttgart, von Klizing offered Schön a permanent position at the prestigious Max Planck Institute he was director of, but Schön refused. Things were going too well at the Bell Labs. In fact, organic crystals continued to work wonders. Schön announced that he has been able to build a laser by interposing a tetracene single crystal between two thin layers of aluminum oxide using the deposition technique available at Konstanz. Actually, this result began to ring bells with some colleagues. One of them was Federico Capasso, an Italian scientist who had a prominent position at Bell Labs, and is one of the world's leading specialists in laser technology.

Some aspects of Schön's measurements were far from convincing, so new experiments were requested to confirm or refute that extraordinary result. Oddly enough Schön himself took care of these new measurements, going to Konstanz and coming back with reported data that, without completely waiving all concern, seemed to reassure the skeptics. Eventually the results passed internal audits, and were considered fit for publication, although something remained unexplained. When faced with new physical phenomena, you cannot expect to understand everything right away! A paper was promptly delivered to *Science* in the summer of 2000, and was accepted in a three-week record time, when usually at least three months pass between submission and acceptance.

One might question the appropriateness of publishing studies in which some aspects are not entirely convincing, but this risk is inherent in today's science. Waiting until all details are clarified can take a long time, and somebody else can claim priority on the discovery. The policy of journals such as *Science* and *Nature* is to publish studies that open new paths and stimulate original research in new directions, although this may require the inclusion of preliminary and perhaps not fully understood data, with all the dangers that this implies. Schön's entire strategy had dwelled on this ambiguity. Moreover, the moment could not have been more propitious for the miracles of electronics based on plastic materials. In October 2000, the Nobel Committee announced the Physics Prize was to be awarded to Alan Heeger, Alan MacDiarmid, and Hideki Shirikawa. The threesome had discovered the first conducting polymers in the 1970s, demonstrating that plastic too (if only under certain conditions) can be an electrical conductor, like the copper of our electrical wires. For those who work in the field that's an apotheosis, such that the magazine *Physics Web* created the expression 'plastic fantastic'. We seemed to be witnessing a revolution, and Schön was the ultimate protagonist.

An exciting ride!

Obviously, a sprinting race like Schön's progress could not be done without colleagues from other labs trying to reproduce the results and share data, information, and materials. Some researchers asked Schön for a sample of his material to better characterize it. Especially, his thin layers of aluminum oxide prompted curiosity (and envy) given the difficulty of obtaining them. David Muller, a colleague at Bell Labs, explicitly asked Schön to provide him with samples to make some measurements. Schön agreed and promised to prepare them on his next trip to Konstanz, but then, once he returned, he said he had forgotten. If the circumstances today seem suspicious to us, it was not really so for Muller. 'He does not accept intrusions', so Schön's Bell Labs colleague thought, or he did not want to share his business with others or be forced to accept collaborations. With such a success as Schön was having, obviously many people wanted to jump on the bandwagon and work with him. So: no samples and no independent checks.

In afterthought, a more strict policy of internal result sharing at Bell Labs would have avoided the catastrophe. Someone also started to press Schön to let people better understand how his 'magic' deposition machine in Konstanz worked. In fact, repeated attempts to get the same samples from other sources invariably failed. But there were many details to check, and it was always possible some ingredient was missing. Many interested people started posing very straightforward questions to Schön, receiving responses that, when and if they arrived, did not bring much light to their inquiries. The results remained unattainable, much to the chagrin of those who had been involved in the attempts: in some extreme cases, deeply frustrated scientists decided to abandon the world of research, feeling unsuitable and unable to reproduce the already published data.

The 'big bang' was now mature. What was at stake was a result that would turn a career inside out, that would launch a scientist toward the firmament of the Nobel Prize. In May 2001, Schön submitted a paper to *Nature*, claiming the implementation of a single layer of organic molecules as a channel for a transistor. For this he coined the expression 'field effect transistor based on a self-assembled monolayer of molecules'. It was the key step toward the molecular transistor, the Holy Grail of molecular electronics. After the paper passed the first internal screening, *Nature* sent it to the referees, whose comments were not really enthusiastic. Scepticism clearly emerged from their reports. Schön and his collaborators, however, were unabashed and concocted a convincing answer based again, as was to be found out later, on manipulated data. *Nature* published the paper 167 days after receiving the manuscript. The impact was enormous, but the next stage was already on the way. At the end of the year, a paper was sent to *Science* where it reported the first transistor based not on a molecular layer, but *on a single molecule*! This is the extreme limit that nanotechnologies strive to attain, the creation of devices based on individual molecules. The path toward an enormous increase in computing power at infinitesimal costs seemed open. The echo hit the media around the planet. Many labs around the world tried to replicate the experiment. And thus, a host of young researchers set to work day and night trying to build the fantastic devices described by Schön in his paper, resulting only in frustration and discomfort after a long sequence of failures. But the fact of being caught in a fraud was so far from everybody's mind that the common belief was that some essential information was lacking, that the samples were not as pure as they should be, that some elementary step had been missed; in short, there could be a thousand and one reasons why reproduction did not succeed, foremost, the inadequacy of other scientists to prepare the new device.

'Annus mirabilis'

And so we come to 2001, annus mirabilis. This year Schön publishes four papers in *Nature* and as many in *Science*. On average, he coauthored one paper every nine days. His success spanned all fields of physics, including new materials exhibiting superconducting behaviour (superconductors are substances that in particular circumstances show zero resistance to the flow of electric current). He initially succeeded with an organic polymer, polytiophene, and then with a completely different material, a mixed copper and calcium oxide. Not surprisingly, Schön's awards were countless: along with two collaborators, Kloc and Batlogg, he received the prestigious Braunschweig Award, and was selected for the Young Researcher Award of the American Material Research Society. Someone starts talking openly about a Nobel Prize. It was, however, just within Bell Labs that some doubt began to circulate.

At an internal seminar, held on 31 October 2001, Schön spoke about the extraordinary discovery of the single-molecule transistor. Questions and protest from the floor put such a pressure on the speaker that Federico Capasso, who chaired the presentation, had to intervene to quiet down the public so as to let the speaker finish his talk. Bombarded with doubts and questions, Schön, unperturbed, always answered in the same way: 'This is what I measure', 'This is what I see', leaving others with the burden of explaining his extraordinary but increasingly unlikely results. The idea that Schön might be cheating began to circulate, but the enormity of such a fact prevented many from giving an unequivocal voice to their doubts. More and more often Schön's answers to questions were vague; he sometimes even contradicted himself. It couldn't be absolutely excluded that Schön may have forgotten some details of his own work, but the suspicion that he was a plain liar gained credit.

People started looking at already published data and results with a different eye. And so, by examining a set of data produced by Schön and their statistical distribution, a Bell Labs researcher, Don Monroe, realized that the linear fitting of these data were too good to be true, as if someone had removed more or less consciously some of them, or had adjusted the associated experimental error. Monroe came to a shocking conclusion: the only way to explain these data was that they had been created artificially. With this conviction Monroe decided to share his doubts with Capasso, but, out of fairness, he also copied Schön into an e-mail to inform him. Schön replied that he could not rule out the possibility that some data may have been unintentionally omitted and promised to reanalyse the raw data. At that point, Monroe, with a quite explicit email whose subject was 'smoking gun', exposed the situation to the Bell Labs executives, alleging that on a statistical basis there was a 90 per cent probability that the data had been distorted by human intervention, with a 50 per cent probability that it had been intentional.

All true?

Of course, the misgivings were not just internal to Bell Labs. On 23 November 2001, Schön received a note from *Nature*, letting him know that Paul Solomon, an IBM Yorktown Heights scientist, had just submitted a letter in which he explicitly criticized the result of the single-molecule transistor. The letter concluded that the work contained several critical points not properly addressed by the authors, which cast doubts on its reliability. As is customary in such cases, the journal offered Schön and his coauthors the opportunity of counter-deduction. In their reply, they said that they had never claimed to have built a single-molecule transistor, but simply to have made some very important steps in that direction, and that obviously further work was needed to confirm the whole thing. Solomon's critical letter and Schön's response were sent to external reviewers who concluded that Solomon's doubts had already been partly highlighted during the review process of the original paper. According to these referees, there was nothing particularly new in Solomon's contention, and thus *Nature*'s editors suggested that he publish his comments and results in another specialized journal.

But from now on, scepticism spread. Among those who decided to look more closely at the set of findings published by Schön during the previous months was Lydia Sohn, whom we met at the beginning of this chapter. She was the first to notice that, although apparently different, the data of the experiments published in the two papers of *Nature* and *Science* on the molecular transistor were actually identical. And it was Lydia Sohn who reported the issue to Liesbeth Venema, making for the first time an allegation of fraud against Schön. After recovering from her astonishment, Venema consulted with other editors of the magazine who decided to contact Schön for clarification. More and more on the defensive, he admitted that in fact, yes, the data were the same, but it had been a mere factual mistake made when submitting the figure of the *Science* article. The data published in *Nature*, he maintained, were correct. But the whole story became more and more fishy.

The coup de grace arrived in those days from Paul McEuen, a professor at Cornell University. Analysing the data published by Schön, McEuen made a strange observation. Whenever a physical experiment is performed, there is an intrinsic effect known as 'background noise', hard to remove, due to random oscillations during measurement. Believe it or not, the background noise shown in one of the figures published by Schön was identical to that previously reported in a figure referring to a completely different experiment. It is extremely unlikely, not to say impossible, that two different experiments will have the same background noise. It was the irrefutable proof that the data were generated on a computer by post-processing, and did not come from direct measurement. No less than six cases of obvious duplication in five different papers were eventually found.

Without notifying Schön, McEuen got in touch with Federico Capasso, warning him that there was a big problem. It was the end of the show.

'Game over'

In May 2002, Bell Labs appointed an inquiry committee. In addition to Schön, his closest coauthors were also under scrutiny: Zhenan Bao, Bertram Batlogg, and Christian Kloc. It turned out that no one except Schön had ever had direct access to the original experimental data. Schön pleaded not guilty and claimed to have possibly swapped files, but to never have altered or, worse, faked the measurements. The committee noted with dismay that for most of the published works no original data were available, nor was there any laboratory notebook where experiments were described and recorded. Schön, incredibly, declared that he has deleted all the data to save space on his computer's hard drive. It also turned out that the original samples of his devices no longer existed since, according to Schön, they were destroyed or dispersed. On 25 September 2002, the committee concluded his examination by sending to Bell Labs management a final document, which was, at the same time, open for public perusal. The document listed 24 cases of misconduct and reported irrefutable evidence of manipulation or duplication in 16 of them. Schön was sacked on the spot and, in the company of two security guards, driven out of the building into which he will never set foot again.

The aftermath was painful. In the weeks and months that followed, the journals that had published his papers began internal inquiries to verify all that had been reported. Eventually, more than 30 published papers were found to be faked and marked for withdrawal, amongst which 7 were in *Nature* and 9 in *Science*. Assigned prizes were revoked and a refund of the awarded money was asked. In June 2004 the University of Konstanz, despite there being no evidence of alteration in his thesis data, withdrew Schön's PhD. The story came to an end, his scientific career was over, and the whole affair left deep, diffused wounds that will perhaps never heal.

Whose fault was it?

As you can imagine, Hendrik Schön's case was at the centre of an intense debate. It has long been wondered how such a massive series of manipulated data could have sailed through the checklists of important journals and been considered real and trustable, so that for some time Schön was offered chairs at some of the most prestigious institutions around the world. Schön was not a criminal mind. He had been fragile and helpless in front of the strong competition and pressure mechanisms that dominate today's modern science. He found himself in the company of some collaborators—first of all his supervisor, Bertram Batlogg—who instead of asking for robust independent verifications of the data that the brilliant young man was producing at incredible pace, acquiesced in trusting their declared truthfulness. Schön had been projected into a system where a former success supports the next one, such that upon seeing the first papers published in journals of enormous impact, the referees of successive papers were influenced and psychologically conditioned. Actually, not everyone in the field had greeted Schön's results with confidence and enthusiasm.

In those years, an Italian professor at Princeton, Giacinto Scoles, visited our department in Milan. In a seminar, Scoles presented his findings on the growth of ordered films of organic molecules on a silicon support, the same problem Schön was investigating. At some point in the presentation, Scoles openly criticized Schön's data, citing them explicitly and saying he did not consider it possible that molecules could organize themselves as Schön was describing. At that time I had never heard that name, and I was amazed at such a direct and bold attack on a colleague in his absence. Only afterwards did I realize its meaning. After the fraud was discovered, I had the opportunity to speak to Scoles about the whole story. He told me that he had been consulted on several occasions as a referee of Schön's papers, and that in many cases he had recommended rejection. But other reviewers had been more indulgent, and eager to be the first to publish results that could mark the history of science, even very demanding journals had been lured into compliance.

A subtler but relevant aspect of Schön's work is that it was not altogether outlandish. It was a sort of anticipation of realistic outcomes, although no one knew when they would materialize. Sniffing astutely around, he offered sensible results to a community that eagerly awaited them. His strategy was to outline a path to things that could actually be accomplished some day, but just with a time lag that would prevent them from claiming priority. Schön, however, did not seem to realize that he could never get away with such serial cheating, so how and why all this happened remains a mystery. Surely the very strong pressure experienced at Bell Labs was one of the triggering factors. As outlined earlier, this problem is more widespread than one may think, and it poses a very acute and dangerous threat.

Just a few rotten apples?

How diffuse is scientific fraud? Is it possible to estimate its impact? The question is obviously crucial and the answer is a subject of debate. In general, cases of fraudulent conduct are few and the protagonists are considered isolated cases of 'rotten apples'. According to some studies, however, what comes to light is just the tip of an iceberg and that there are many other cases that have never been discovered.[71] The debate then focused on how to define scientific bad practice and how to identify a misconduct.[72] Schön's story is a pure fabrication of never measured data, but there must also be several ways to distort measured data, along with the already discussed phenomenon of plagiarism. The latter is paradoxically considered less damaging than fraud (though certainly more widespread) as it at least does not affect adversely scientific knowledge.

Improper data treatment can take very subtle forms, like biased selection and interpretation to prepare a set that statistically best supports a given hypothesis. It is enough to selectively report the data consistent with a certain interpretative model, discarding and not commenting on others. There is therefore a wide range of misbehaviour, from simple superficiality and negligence to the improper handling or full misuse of obtained data. The consequences, of course, can be disconcerting.[73] These behaviours are therefore classified as 'questionable research practices', and the borderline with real fraud becomes very fuzzy. It should be recalled that erroneous interpretation or perhaps a too optimistic reading of some results belongs to the normal category of human fallacy, without having to be classified as misdemeanor proper.

Some surveys have tried to quantify the frequency of incorrect practices. According to one, fraud involves one out of every 100,000 scientists;[74] another estimate reckons one in every 10,000.[75] Considering the number of withdrawals of published papers in the PubMed Medical Database, the estimated frequency is 0.02 per cent, which would lead to infer that such a fraction of scientific work is fraudulent.[76] In recent years various surveys have been conducted by questioning scientists directly. It turned out that cases of misconduct, including plagiarism, are around 1–2 per cent of the total (see also endnote 74).[77] One study concluded that, on average, 2 per cent of respondents admitted to having falsified data at least once in their career. These are indeed low percentages and small numbers, but not negligible, considering the total number of papers published each year, and, sadly, these numbers are increasing. According to a 2011 *Nature* survey,[78] the number of scientific papers being withdrawn has seen an upsurge, something like 1,200 per cent in the past ten years, while the number of published studies has increased 'only' by 44 per cent. Obviously, not all the papers that have been withdrawn are due to some kind of cheating, which is, however, estimated at about half of the cases.

The fact that the problem has become a serious one is highlighted by the attention it received from the prestigious Alexander von Humboldt Foundation, one of the most famous in the world, with its over 26,000 carefully selected members from 130 countries, among which about 50 Nobel Prize winners. The reason why this historic and distinguished organization had to deal with the matter is that at the beginning of 2015 Jens Förster, a social psychologist of the Ruhr-Universität in Bochum, had to return funding worth 5 million euros to the foundation after his university discovered that some of his research results had been doctored.[79] Even a historical institution of excellence, with a long tradition like that of the von Humboldt Foundation, is no longer immune from contagion!

8

Do we still believe in science?

When in March 2016 Robert De Niro invited Andrew Wakefield to present his documentary *Vaxxed: From Cover-up to Catastrophe* at the Tribeca Film Festival in New York, he certainly did not imagine he would raise a planetary controversy: indeed, he was forced to step quickly back and to withdraw the invitation, with so many public apologies. De Niro has an autistic son, Elliot, born in 1998, and his intention was to stimulate the discussion of how this mysterious pathology can still arise. Nobody questions the fact that there should be open discussion on such controversial topics. But the fundamental premise is that everything should be based on factual and objective data, barring inventions, misinformation, or worse, bad faith. This is the case with Dr Andrew Wakefield, whose 'contribution' to science and society sparked the above-mentioned controversy that has spread all around the world.

Born in 1957 and active in England as a physician and surgeon, Wakefield obtained public visibility with wide media exposure in 1998 when he published a disruptive scientific paper in the medical journal *The Lancet* in which he argued that the trivalent vaccine for measles, mumps, and rubella is responsible for the insurgence of autism and intestinal diseases, generating fear and alarm in public opinion.[80] Over the following years, however, other researchers failed to reproduce Wakefield's results and to confirm his hypothesis about the existence of a correlation between vaccines and autism or gastrointestinal diseases.[81] In 2004, moreover,

The Overproduction of Truth. Gianfranco Pacchioni.

DOI: 10.1093/oso/9780198799887.001.0001

a *Sunday Times* reporter, Brian Deer, discovered the existence of a heavy conflict of interest (never disclosed by Wakefield) and found other questionable practices as the basis of his studies. The story had a strong impact, to the extent that most of Wakefield's coauthors decided to withdraw their support to the work.[82] The British General Medical Council opened an investigation that highlighted numerous anomalies and reckless aspects in the work behind the 1998 paper, including the fact that autistic children had undergone invasive medical procedures without their consent.

In January 2010, the General Medical Council concluded that Wakefield had violated dozens of ethical rules, and had operated dishonestly and irresponsibly.[83] At the same time, *The Lancet* withdrew the work in question, reporting in an editorial that the data had been entirely falsified.[84] It was eventually discovered that Wakefield had been bribed to alter the results in order to support a series of lawsuits filed by an attorney against vaccine-producing pharmaceutical companies.[85] Wakefield had even patented an alternative, pretend vaccination procedure to replace the trivalent vaccine he had charged with causing autism. In short, it was a wholesome collection of scientific malpractices in a single stroke. Wakefield was ejected from medical associations, and further investigations have only confirmed his dishonest and fraudulent conduct. But the damage has been done: the UK, the USA, and other countries where Wakefield's work has had a remarkable echo have witnessed a fall in vaccination percentages over the years, significantly lowering the population's immunization level, causing increased number of measles, some serious illnesses, and even a few deaths. Against all evidence, Wakefield has consistently denied any misconduct and reiterated his message against vaccines, obviously finding plenty of web support for his claim to be a victim of conspiracy. In this respect the Web has acted as a powerful and filter-free loudspeaker. Despite the fact that the highly reputed scientific world has been unyielding and unanimous against Wakefield, his 'theories' survive and freely ride the Internet.

There have been other, hair-raising consequences of the case. In June 2012 an Italian Court in Rimini ruled that trivalent vaccination was the cause of autism in a 15-year-old boy. In making that judgment, the court made reference to Wakefield's paper in *The Lancet*, ignoring what the entire scientific community had brought to light in the years following the paper's publication. Fortunately, in the appeal trial the sentence was overturned by the Bologna Court. What to say? The story warns of the dangers of such a chain of events.

All this brings us to the heart of the problem: the relationships between science and society. Hendrik Schön's story, told in the previous chapter, shook the foundations of the validation system of scientific results. As repeatedly stressed, such lies in science cannot go far, although Schön's comedy lasted almost three years (not a short period, anyway). Once the cheat is discovered, the system corrects the mistakes, the wounds cicatrize, and the whole episode enters without too much damage into the list of proverbial scientific mishaps, mainly food for public curiosity. But Wakefield's case is different. Clinical work requires a long time, with complex and costly investigations, to be reconsidered; the first results may not be sufficient to fully clarify the issue and therefore further verifications may be required. In the meantime, unscrupulous and irresponsible use of some conclusions spreads damage and can wreak havoc with social standards, affecting innocent victims. Due to a peculiar but common psychological mechanism, the statement that something is harmful has an infinitely stronger impact on the media than the statement that something is harmless. Negative warnings foster doubts, generate suspicion, find their way through deeper psychological mechanisms, and stick there almost indelibly. Not even the highest authority succeeds in wiping out the fears sparked by false prophets who spread the seed of doubt.

False prophets, false alarms

Another enlightening example is the fake news about so-called chemical trails, according to which the white traces released by flying aircrafts are not water-vapour condensation trails but streaks of chemicals or biological substances deliberately dispersed in flight for various purposes. For those who want to find support for these ideas, it is enough to read an article by J. Marvin Herndon published in the *International Journal of Environmental Research and Public Health* in 2015.[86] You should not be impressed by the name of the journal: it does not enjoy a particularly good reputation.[87] Herndon uses this article to explain that aircraft tanks spill a toxic substance into the atmosphere apparently with the intent of causing climate change. He claims that governments and military apparatuses of Western countries are behind the chemical trails phenomenon. His descriptions and conclusions are totally unwarranted, and any person with a grain of common sense should wonder how it was possible that such an article could overcome the filter of the (presumed) peer-review process. Meanwhile, a website republished the article with a frightening announcement: 'Exceptional: Scientific work with peer review confirms the presence of ashes in chemical trails'. There is no need to emphasize the effect that this has on the public. Small positive note: some counter-reaction must have occurred, because the paper was then withdrawn. But don't be too optimistic. Many will insist that Herndon's words are pure gold.

The validation of results of scientific research thus plays a fundamental role. Unless press releases are accompanied by a rigorous verification process, scientific knowledge is not disseminated through press conferences. Until recently the peer-review process, entrusted to the scientific community, has been (with just a few flaws) robust enough to check and support factual evidence. But all of that has been discussed so far in this book; in particular, the unconditional proliferation of researchers, scientific journals, and ensuing publications and diffusion has progressively weakened the authority of scientific affirmation, making it less and less credible, more and more subjective. All this points to an immense problem in the relationship between science and society.

Science and democracy

Science is not democratic, not in the sense that we usually attribute to this term. And it does not have to be, even though people tend not to be aware of that. There are no public polls to decide whether the second principle of thermodynamics or the theory of relativity is valid. Only experimental verification can provide evidence of the rigor and wide applicability of a theory and confirm its formal structure. Things become more complex when it comes to questionable issues, those where even scientists can hold widely different views. The key point is to provide scientifically consolidated and scientifically substantiated data to support one's opinions. The difficulties often stand not so much in the interpretation of present data but in their extrapolation to future scenarios.

Extrapolation is a complex and dangerous practice, which can lead to conclusions very far from reality. Science and scientists, however, may not necessarily provide definitive answers, but rather solid data upon which citizens are called upon to make choices, possibly once an informed opinion has been developed. To achieve a fair level of awareness and understanding, a scientist needs years of hard work, study, and discussion. Inevitably, this makes science an elitist enterprise, often out of tune with the perception of common people. And common people tend to distrust what they do not understand.

The process of forming opinions in science is totally different from what proceeds in the world of the press. Faced with a political or social problem, a journalist interviews two would-be experts from opposite sides of the issue, and leaves the reader (or viewer) with the burden of forming an opinion, once the two bells have been heard. Applied to science, this way of proceeding can produce perverse effects. When it is pretended that the opinion of rambling imposters without specific expertise has the same resonance as that of a recognized specialist in the field who has years of accumulated experience and a wealth of scientific production, no democracy accrues—only damage. In disseminating scientific knowledge, a journalist is called to a difficult but fundamental exercise: to assess the credibility of his speakers. Today this is getting more and more difficult. Indeed, often the 'official' scientist is seen as the bearer of strong and hidden interests, as the representative of some consolidated power, while the occasional talking head assumes the role of a heroic fighter against the establishment, of an untamed David standing alone against the mighty Goliath.

The lack of trust in science and in scientists, or just the confusion generated by the excess of information and the arrogance of pseudo-scientists and scientific pseudo-publications, are recent developments extremely dangerous for our society and for our democracy. On the one hand, no one expects the public to have the full expertise necessary to understand the intimate nature of scientific issues; on the other hand, a more widespread scientific culture and awareness, and a better perception of the role of science for the growth of society, are highly desirable. The global problems that emerge in a planet nowadays overpopulated and with limited resources can only be solved with the help of expert scientists, most likely the only ones who can propose reasonable solutions. Science, however, is not a religion, although some highbrow scientists sometimes look like ministers engaged in the fierce propagation of their 'faith'. A good scientist should never forget that a modicum of scepticism is always indispensable. Nevertheless, whenever all information is needed before making a decision for all of society, it is only the scientific community that should be addressed. Failing to do so will enable severely wrong decisions to be made, often with catastrophic consequences. Even worse outcomes can appear when trying to twist science's arm to allow all sorts of improper conditioning, like the will of political power or the whims of public pressure.

With the precise aim of releasing scientific knowledge from all social and political constraints, at the beginning of the seventeen century the first European scientific societies and academies were created. Their purpose was to promote scientific methodology, to provide authoritative certification of the results, and to represent a clear reference point for anyone called upon to make decisions on a given matter. While not perfect, the system, based on the representativeness and authority of the scientific community, has worked well and been the basis of the great scientific and technological advances of the past two centuries. Slowly, but progressively, this system began to crack. In the first place, scientific societies have not always been waterproof to the pressure of political power (which is often the source of their financial support). Moreover, the appearance of many forms of scientific communication that are totally out of touch with the consolidated rites and procedures of these societies makes their role more and more opaque. If we add to this the problems arising from the inordinate growth of scientific production, as we have outlined in previous chapters, the danger of missing solid reference points, of mixing in a haphazard way good and bad information, valid and erratic concepts, and sound and honest scientists and perfect charlatans, becomes evident. The social cost of all this could be enormous.

Where does science go?

At this point, the meaning of the question that gives this book its title, 'The Overproduction of Truth', should be more clear. But we must recognize that very similar problems are affecting many other sectors of modern society. For example, what about an economic system dominated by financial markets increasingly at the mercy of emotional and irrational reactions, of speculative bubbles, and of ethically reproachable behaviours where the economic interest of a very few determines the social destiny of entire nations? And what about interpersonal relationships, now largely subject to the stringent laws of social media, where the rule is the universal violation of your very intimacy, with all the follow-up of media gossip, online insults, persecution and verbal group aggression behind the screen of anonymity? Or a political milieu becoming less and less attentive to real and long-term problems, and increasingly intent on catching up with deeper popular feelings, dramatizing dangers, stimulating the worst fears, in order to gain low-cost consensus? In a world evolving in such directions, one cannot expect science to remain a happy and unspoilt island, a sort of fairyland totally immune from such regressive trends.

Ultimately, science is made by flesh-and-bone women and men, with all the problems, insecurities, and typical frailties of ordinary people. However, and despite all the problems we have discussed, from this point of view we can say that the world of science is and remains an essentially objective, sound system with lively antibodies and ethical principles that are still largely rooted in the majority of its practitioners. So we are not at the last resort, at the point of no return: definitely not. But the symptoms emerging more and more frequently, as sketched out in the previous chapters, must be recognized and analysed, and then translated into actions that can restore the focus on curiosity, dedication, and even passion, all essential elements of that gratifying intellectual enterprise: scientific research.

What to do then? First of all, there is a need to change the parameters that measure scientific productivity, shifting their loadings more and more from quantity to quality. Provocatively, it has been proposed that every person who obtains a PhD be given a fixed number of tokens, say 100, to be used each time that person intends to publish a paper. That number would represent the highest possible scientific productivity of a researcher in his or her career, so that one should think twice before wasting one of those precious tokens on a lacklustre publication. Others have suggested that each researcher can publish annually only a given number of papers. These are, as you may well realize, provocative and unrealistic proposals that nevertheless point a finger at the problem. Of course, much must be done also at educating doctoral students and young researchers to carefully plan their careers and to focus always on research and investigations of high quality and relevance.

It is not a question of good or bad behaviour; it is a question of convenience and maximizing true scientific progress and output. This translates into the balance between emphasizing quality and emphasizing quantity in scientific research. As observed in a recent study by Edwards and Roy,[88] a process that overemphasizes quality may require long times to validate, replicate, and confirm results. This would minimize mistakes but would also lead to very slow progress due to overcaution. At the other extreme, emphasis on quantity implies less scientific rigor, resulting in a large number of low-quality publications. In the end, this also produces little true progress. An optimum lies somewhere in between, as shown in the curve of Figure 7. However, there is another important aspect to consider. Under extreme pressure

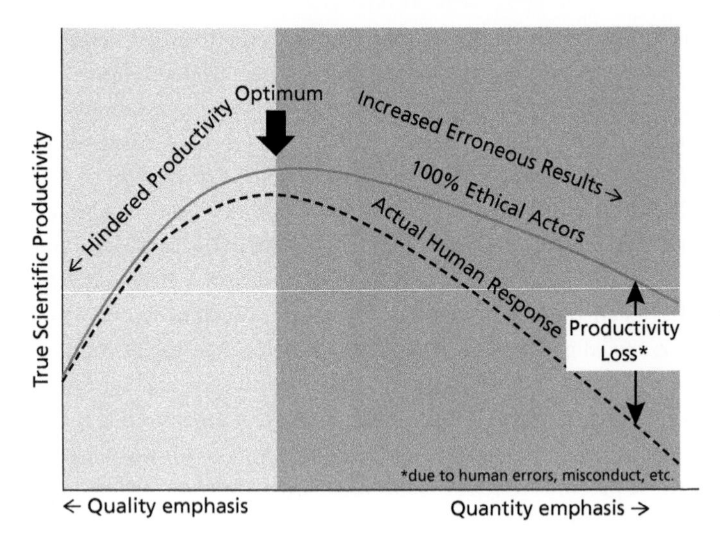

Figure 7 True scientific productivity versus emphasis on research quality or quantity.

Source: From Edwards, M. A., Roy, S. (2017), Academic research in the 21st century: maintaining scientific integrity in a climate of perverse incentives and hypercompetition, *Environmental Engineering Science* **34**, 51–61.

and hyper-competition, if the emphasis is on quantity, as it is now, the problem is exacerbated by cases of misconducts and lack of scientific integrity. Productivity and progress can be severely affected by massive numbers of incorrect, irreproducible, or simply irrelevant studies. Of course, this is not an issue when emphasis is on quality, as there is no special gain in following unethical or superficial scientific practices. Right now, it seems that we are moving towards the right side of the curve. Emphasizing quality over quantity should be one of the targets of the scientific community in the near future.

The need is also felt for improving and strengthening the common perception of science, through a continuing dialogue between researcher and society. We need to better communicate what a scientist's work involves, openly underscoring the inherent risks inherent to research, neither concealing nor dramatizing them. There is definitely a need to point out that a modern scientist is not the eccentric, isolated, dishevelled operator as in people's imagination, working in a laboratory with steaming ampoules, every now and then delivering some genial idea like a novel Newton under his apple tree. Quite the opposite: nowadays researchers work in teams of ever-growing size, spending years of endless patience on the topics that catch their attention, and investing massive amounts of dedication, perseverance, determination, commitment for the achievement of their goals. As aptly stated by Aaron Klug, a Nobel Prize winner for Chemistry for his studies on the structure of nucleic acids and proteins:[89] 'The greatest illumination in scientific research comes from people who have the patience to develop an intimate understanding of the problem, who have the space and the freedom to take on professional risks, and who know how to make creative use of the surprises they encounter when they do so.'

But all the above may not be enough. It is not enough to improve the ways of communication between science and society, and it is not enough to apply corrective action on evaluation mechanisms. The picture of the modern scientist is still positive, but it is likely to be altered by current, adverse events: extreme competition, the need to publish more and more quickly, the superficiality with which problems are considered, and the reduced times available for the verification and appropriation of the results, one's own as well as those of others. Besides, there is a concrete danger of weakening the ethical principles that must inform the practices of scientific operators. Briefly, the solution to the problems of contemporary science can only come from its main actors, the scientists themselves.

The ethics of science

Throughout the past century, science has made reference, more or less consciously, to some ethical principles. Four of these were clearly formulated by the famous sociologist Robert Merton in 1942[90]: universalism, communalism, disinterestedness, and organized scepticism. The concept of universalism is to indicate that anyone can contribute to science irrespective of race, nationality, culture, religion, or gender. The second concept, communalism, can be translated into collegiality: science is not a private matter, but a shared property. All scientists should have equal access to scientific knowledge and there should be a sense of shared ownership behind the very concept of collaboration. Confidentiality, the secrecy of data, is in fact the opposite of the concept of communalism. Clearly, if all research results were to remain secret, scientific and technological development would be terribly slow.

The third principle, disinterest, may sound a bit naïve and out of fashion in the contemporary world where almost everything seems to drive to an immediate and tangible benefit. Merton felt that researchers should find their motivation in the pleasure of discovery, in the understanding of new phenomena, and generally speaking, in the advancement of knowledge. The scientist's chief motivation should therefore be not in career advancement, nor in the quest for awards and recognitions, nor in wringing out large funding per se. These are possibly to be considered as due consequences, but should not be the primary purpose of the scientist's activity. This is a 'purist' vision, but it has its roots in a very important principle: science as a vocation, rather than just as a business. And there is at least a grain of truth in this: for example, what other category of professionals shares its wisdom for free, like when it comes to act as reviewer of papers or projects, to provide opinions, or to give lectures in public? To the best of my knowledge, there aren't many others.

A very important, real pillar is the fourth principle formulated by Merton: organized scepticism. The scientific community must always verify the accuracy and reliability of every statement and of every result. Without this, there is no progress in science, because it is only on such premises that unreliable results and inaccurate or contradictory propositions can be put aside.

Today one may wonder whether the moral principles formulated by Merton a long time ago are still valid, and if it still makes sense to speak about science ethics. Certainly, some of these principles, which have been among the prerequisites of the extraordinary scientific development of the past century, are now under severe strain.

Let science be slow, but not too slow

Scientific research, original and free from constraints, remains an essential tool for addressing the global challenges faced by the planet (food and water shortage, booming population against limited resources, energy supply, aging society, pollution control, urbanization, and so on). And, as already mentioned, science remains a wonderful intellectual adventure. If done properly and successfully, it is probably the most rewarding of all human activities. And this is still the foundation and creed of most researchers, certainly of the best ones. Today, some advocate returning to the practices of old times; in short, they would like to do science as I knew it at the beginning of my career. Some old-timers have even started a trend in the wake of the so-called 'Slow Food' movement (an association of traditionalist Italian gourmets hating fast food) calling it, by analogy, 'Slow Science'. Does this make sense?

Certainly not for those who are starting now a scientific career, having to commit all their energies to emerge and not be smashed out by competition. Only mature and experienced scientists should consider such issues, opening a discussion and outlining a new character, the responsible scientist: first of all, responsible towards society, which invests its resources in science and to which scientists should return a potential increase in general knowledge and positive spinoffs of their own research; responsible towards colleagues, which means abiding by a set of basic ethical rules; responsible towards younger researchers, towards students, that is, those who will someday carry on the banner. But above all, perhaps we must seriously consider a reduction in the number of researchers in basic science, having them more motivated and truly inspired and selected by better criteria. While waiting for something like this to happen, it is not without advantage to conclude this book with the Slow Science manifesto published in 2010 by a group of German professors:[91]

> We are scientists. We don't blog. We don't twitter. We take our time. Don't get us wrong—we do say yes to the accelerated science of the early 21st century. We say yes to the constant flow of peer-review journal publications and their impact; we say yes to science blogs and media & PR necessities; we say yes to increasing specialization and diversification in all disciplines. We also say yes to research feeding back into health care and future prosperity. All of us are in this game, too. However, we maintain that this cannot be all. Science needs time to think. Science needs time to read, and time to fail. Science does not always know what it might be at right now. Science develops unsteadily, with jerky moves and unpredictable leaps forward—at the same time, however, it creeps about on a very slow time scale, for which there must be room and to which justice must be done. Slow science was pretty much the only conceivable science for hundreds of years; today, we argue, it deserves revival and needs protection. Society should give scientists the time they need, but more importantly, scientists must take their time. We do need time to think. We do need time to digest. We do need time to misunderstand each other, especially when fostering lost dialogue between humanities and natural sciences. We cannot continuously tell you what our

> science means; what it will be good for; because we simply don't know yet. Science needs time.

It may look a bit naïve, but one could hardly deny that the manifesto touches a very sensitive spot. Of course, too much is too much. The magazine *Scientific American* reported the case of a paper submitted almost half a century after being commissioned.[92] The author, Ian Shine, was a physician working on the island of Saint Helen, in the middle of the Atlantic Ocean, in the early 1960s. There he carried out clinical and genetic studies on a population that lives in substantial isolation. Recently, Shine contacted the journal, as follows: 'I have no idea how long contracts are valid but I hope you will consider for publication the enclosed response to your kind offer albeit somewhat delayed. My perspective as a slow writer is that publishers are too impatient, and if they waited a little while—in this case 47 years—they would receive a worthwhile submission.'

We should definitely find a decent compromise between today's instant publications and Shine's 47-year enterprise.

Endnotes

1 Cantelli, G. (ed.) (1958), *La disputa Leibniz-Newton sull'analisi*, p. 200–2, 206–7 (Boringhieri; in Italian).

2 Geva, T. (2006), Magnetic resonance imaging: historical perspective, *Journal of Cardiovascular Magnetic Resonance* **8**, 573–80.

3 Rinck, P. A. (2016), The history of MRI. In: *Magnetic Resonance in Medicine*, 9th edn. Available at; http://magnetic-resonance.org/ch/20-04.html, accessed October 2017.

4 Pacchioni, G. (2000), Quantum chemistry of oxide surfaces: from CO chemisorption to the identification of structure and nature of point defects on MgO, *Surface Review and Letters* 7, 277–306.

5 de Solla Price, D. J. (1963), *Little Science, Big Science* (New York: Columbia University Press).

6 Various authors (2016), *Science and Engineering Indicators 2016* (Washington, DC: National Science Board, National Science Foundation). Available at http://www.nsf.gov/statistics/2016/nsb20161/#/, accessed October 2017.

7 Plume, A., van Weijen, D. (2014), Publish or perish? The rise of the fractional author . . . , *Research Trends* **38**. Available at https://www.researchtrends.com/issue-38-september-2014/publish-or-perish-the-rise-of-the-fractional-author/, accessed October 2017.

8 Khabsa, M., Giles, C. L. (2014), The number of scholarly documents on the public web, *PLOS One* **9**, e93949.

9 Orduña-Malea, E., Ayllón, J. M., Martín-Martín, A., López-Cózar, E. D. (2014), About the size of Google Scholar: playing the numbers. ArXiv eprint available at http://arxiv.org/abs/1407.6239, accessed October 2017.

10 Weisz, P. B. (1968), Exponentials, *Advances in Catalysis* **18**, 7.

11 Bormann, L., Mutz, R. (2015), Growth rates of modern science: a bibliometric analysis based on the number of publications and cited references, *Journal of the Association for Information Science and Technology* **66**, 2215.

12 Various authors (2012), *Science and Engineering Indicators 2012* (Washington, DC: National Science Board, National Science Foundation).

13 Various authors (2011), *Knowledge, Networks and Nations: global Scientific Collaboration in the 21st century* (London: Royal Society).
14 Cronin, B. (2001), Hyperauthorship: a postmodern perversion or evidence of a structural shift in scholarly communication practices? *Journal of the American Society for Information Science and Technology* **52**, 558–69.
15 ATLAS collaboration (2012), Observation of a new particle in the search for the Standard Model Higgs boson with the ATLAS detector at the LHC, *Physics Letters B* **716**, 1.
16 ATLAS collaboration (2012), Observation of a new boson at a mass of 125 GeV with the CMS experiment at the LHC, *Physics Letters B* **716**, 30.
17 King, C. (2012), Multiauthor papers: onward and upward, available at http://archive.sciencewatch.com/newsletter/2012/201207/multiauthor_papers/, accessed October 2017.
18 Aad, G., et al. (2015), Combined measurement of the Higgs boson mass in Pp collisions at 7 and 8 TeV with the ATLAS and CMS experiments, *Physical Review Letters* **114**, 191803.
19 Sarewitz, D. (2016), The pressure to publish pushes down quality, *Nature* **533**, 147.
20 Citron, D. T., Ginsparg, P. (2015), Pattern of text reuse in a scientific corpus, *Proceedings National Academy of Sciences* **112**, 25.
21 Baker, M. (2016), Problematic images found in 4% of biomedical papers, *Nature*. Available at http://www.nature.com/news/problematic-images-found-in-4-of-biomedical-papers-1.19802, accessed October 2017.
22 Bik, E. M., Casadevall, C., Fang, F. C. (2016), The prevalence of inappropriate image duplication in biomedical research publications. Available at http://biorxiv.org/content/biorxiv/early/2016/04/20/49452.full.pdf, accessed October 2017.
23 Ware, M, Monkman, M (2008), Peer review in scholarly journals: perspective of the scholarlycommunity—an international study, *Information Services & Use* **28**, 109–12.
24 Olesen Larsen, P., von Ins, M. (2010), The rate of growth in scientific publication and the decline in coverage provided by Science Citation Index, *Scientometrics* **84**, 575–603.
25 Mabe, M. A., Amin, M. (2001), Growth dynamics of scholarly and scientific journals, *Scientometrics* **51**, 147–62.
26 Mabe, M. (2003), The growth and number of journals, *Serials* **16**(2), 191–7.
27 Ware, M., Mabe, M. (2015), The STM Report (International Association of Scientific, Technical and Medical Publishers).

28 Directory of Open Access Journals: http://www.doaj.org/, accessed October 2017.
29 Beall, J. (2012), Predatory publishers are corrupting open access, *Nature* **489**, 179.
30 Bohannon, J. (2013), Who's afraid of peer review?, *Science* **342**, 60–5.
31 https://scholarlyoa.com/ (now closed).
32 Beall, J. (2017), What I learned from predatory publishers, *Biochemia Medica* **27**, 273–8.
33 Hvistendahl, M. (2013), China's publication bazar, *Science* **342**, 1035–9.
34 Normile, D. (2017), China cracks down on fraud, *Science* **357**, 435.
35 Allison, D. B., et al. (2016), Reproducibiliy: a tragedy of errors, *Nature* **530**, 26.
36 Scott, J. F. (2007), Ferroelectrics go bananas, *Journal of Physics: Condensed Matter* **20**, 021001.
37 Gunther, M. (2015), Meteoric rise of perovskite solar cells under scrutiny over efficiencies, *Chemistry World*, 2 March 2015. Available at http://www.rsc.org/chemistryworld/2015/02/meteoritic-rise-perovskite-solar-cells-under-scrutiny-over-efficiencies, accessed October 2017.
38 Begley, C. G., Ellis, L. M. (2012), Drug development: raise standards for preclinical cancer research, *Nature*, **483**, 531–3.
39 Wadman, M. (2013), NIH mulls rules for validating key results, *Nature*, **500**, 14–16.
40 Freedman, L. P., Cockburn, I. M., Simcoe, T. S. (2015), The economics of reproducibility in preclinical research, *PLOS Biology* **13**, e1002165.
41 Baker, M. (2016), Is there a reproducibility crisis?, *Nature* **533**, 452.
42 Ware, M., Mabe, M. (2012), The STM report: an overview of scientific and scholarly journal publishing (STM: International Association of Scientific, Technical and Medical Publishers), available at http://www.stm-assoc.org/2012_12_11_STM_Report_2012.pdf, accessed October 2017.
43 Larivière, V., Haustein, S., Mongeon, P. (2015), The oligopoly of academic publishers in the digital era, *PLOS One*, doi: 10.1371/journal.pone.0127502, accessed October 2017.
44 Garfield, E. (1955), Citation indexes to science: a new dimension in documentation through association of ideas, *Science* **122**, 108–11.
45 Hohenberg, P., Kohn, W. (1964), Inhomogeneous electron gas, *Physical Review* **136**, B864–71.

46 Garfield, E. (1999), Journal impact factor: a brief review, *Canadian Medical Association Journal* **161**, 979.
47 Sheldrick, G. M. (2008), A short history of SHELX, *Acta Crystallographica A* **64**, 112.
48 Garfield, E. (1979), Is citation analysis a legitimate evaluation tool?, *Scientometrics* **1**, 359.
49 International Society for Scientometrics and Informetrics website. Available at http://issi-society.org/, accessed October 2017.
50 Hoeffel, C. (1998), Journal impact factors, *Allergy* **53**, 1225.
51 Hirsch, J. E. (2005), An index to quantify an individual's scientific research output, *Proceedings National Academy of Sciences* **102**, 16569.
52 Gamow, G. (1966), *Thirty Years That Shocked Physics: The Story of Quantum Theory* (New York: Dover).
53 Society for Neuroscience, Annual meetings attendance 1999–2012, available at http://www.sfn.org/sfn/amstats/amstatsgraph.html, accessed October 2017.
54 Radiological Society of North America, Annual meeting, available at http://www.rsna.org/annual_meeting.aspx, accessed October 2017.
55 Ioannidis, J. P. A. (2014), How to make more published research true, *PLOS Medicine* **11**, e1001747.
56 Royal Society (2011), *Knowledge, Networks and Nations: Global Scientific Collaboration in the 21st Century* (London: Royal Society).
57 Elsevier (2013). International comparative performance of the UK research base, report for UK Department of Business, Innovation and Skills. Available from https://www.elsevier.com/research-intelligence/research-initiatives/BIS2013, accessed October 2017.
58 Various authors (2002), *Frascati Manual: Proposed Standard Practice for Surveys on Research and Experimental Development*, 6th edition (Paris: OECD), available at http://www.oecd.org/sti/inno/frascatimanualproposedstandardpracticeforsurveysonresearchandexperimentaldevelopment6thedition.htm, accessed October 2017.
59 National Science Board (2012), *Science and Engineering Indicators 2012* (Washington, DC: National Science Foundation).
60 Organization for Economic Co-operation and Development (OECD) (2017), R-D personnel by sector of employment and occupation, available at http://stats.oecd.org/Index.aspx?DataSetCode=PERS_OCCUP, accessed October 2017.

61 Van Noorden, R. (2014), China tops Europe in R&D intensity, *Nature* **505**, 144–5.

62 The National Academic Press, Appendix F: Science, technology, and innovation databases and heat map analysis, dataset: science, technology and innovation R&D full time equivalent per country 1995–2013; available at http://www.nap.edu/read/18606/chapter/17, accessed October 2017.

63 Perutz, M. F. (1999), Will biomedicine outgrow support?, *Nature* 399, 299–301.

64 Goodstein, D. (1994), The big crunch, in *NCAR 48 Symposium, Portland, September 19, 1994*: available at http://www.its.caltech.edu/~dg/crunch_art.html, accessed October 2017.

65 Larson, R. C., Ghaffarzadagan, N., Xue, Y. (2014), Too many PhD graduates or too few academic job openings: the basic reproductive number R_0 in academia, *Systems Research and Bahavioral Science* **31**, 745–50.

66 Various authors (2017), Survey of graduate students and postdoctorates in science and engineering, National Science Foundation; available at https://www.nsf.gov/statistics/gradpostdoc/, accessed October 2017.

67 Sauermann, H., Roach, M. (2016), Why pursue the postdoc path?, *Science* **352**, 663.

68 Benderly, B. L. (2010), Does the U.S. produce too many scientists?, *Scientific American*, available at http://www.scientificamerican.com/article/does-the-us-produce-too-m/, accessed October 2017.

69 Reich, E. S. (2009), *Plastic Fantastic: How the Biggest Fraud in Physics Shook the Scientific World* (London: Palgrave Macmillan).

70 Ossicini, S. (2012), *L'universo è fatto di storie non solo di atomi*, Neri Pozza.

71 Sovacool, B. K. (2008), Exploring scientific misconduct: isolated individuals, impure institutions, or an inevitable idiom of modern science?, *Journal of Bioethical Inquiry* **5**, 271–82.

72 National Academies of Sciences, Engineering, and Medicine (2017), *Fostering Integrity in Research* (Washington, DC: The National Academies Press). Available at http://www.nap.edu/21896, accessed October 2017.

73 De Vries, R, Anderson, M. S., Martinson, B. C. (2006), Normal misbehaviour: scientists talk about the ethics of research, *Journal of Empirical Research on Human Research Ethics* **1**, 43–50.

74 Steneck, N. H. (2006), Fostering integrity in research: definitions, current knowledge, and future directions, *Science and Engineering Ethics* **12**, 53–74.

75 Marshall, E. (2000), Scientific misconduct—how prevalent is fraud? That's a million-dollar question, *Science* **290**, 1662.

76 Claxton, L. D. (2005), Scientific authorship, Part 1: A window into scientific fraud?, *Reviews in Mutation Research* **589**, 17–30.

77 Fanelli, D. (2009), How many scientists fabricate and falsify research? A systematic review and meta-analysis of survey data, *PLOS One* **4**, e5738.

78 Van Noorden, R. (2011), Science publishing: the trouble with retractions, *Nature* 478, 26–8.

79 Vitzhum, T. (2015), And lead us not into temptation, *Kosmos* **104**, 13.

80 Wakefield, A. J., et al. (1998), Ileal-lymphoid-nodular hyperplasia, non-specific colitis, and pervasive developmental disorder in children, *The Lancet* **351**, 637–41 (retracted).

81 Black, C., Kaye, J. A., Jick, H. (2002), Relation of childhood gastrointestinal disorders to autism: nested case-control study using data from the UK General Practice Research Database, *BMJ Clinical Research* **325**, 419–21.

82 Deer, B. (2004), Revealed: MMR research scandal, *The Sunday Times (London)*, 22 February 2004.

83 Kumar, S., et al. (2010), Report, General Medical Council, available at http://briandeer.com/solved/gmc-charge-sheet.pdf, accessed October 2017.

84 The Editors (2010), Retraction—Ileal-lymphoid-nodular hyperplasia, non-specific colitis, and pervasive developmental disorder in children, *The Lancet* **375**, 445.

85 Godlee, F., Smith, J., Marcovitch, H. (2011), Wakefield's article linking MMR vaccine and autism was fraudulent, *BMJ Clinical Research* **342**, c7452.

86 Herndon, J. M. (2015), Evidence of Coal-Fly-Ash Toxic Chemical Geoengineering in the Troposphere: Consequences for Public Health, *International Journal Environmental Research Public Health*, **12**, p. 9375–90 (retracted).

87 Gerdol, M. (2016), *More Pseudo-Science from Swiss/Chinese Publisher MDPI, Scholarly Open Access*, http://archive.li/6tNTe (checked on October 2017).

88 Edwards, M. A., Roy, S. (2017), Academic research in the 21st century: maintaining scientific integrity in a climate of perverse incentives and hypercompetition, *Environmental Engineering Science* **34**, 51–61.

89 Rees, M. (2011), *From Here to Infinity: Scientific Horizons* (London: Profile Books).
90 Merton, R. K. (1973), *The Sociology of Science: Theoretical and Empirical Investigations*, 267–78 (Chicago: University Chicago Press).
91 The Slow Science Academy (2010), The Slow Science Manifesto. Available at http://slow-science.org/; accessed October 2017.
92 Rusting, R. (2016), A manuscript 47 years in the making, *Scientific American*, available at http://blogs.scientificamerican.com/at-scientific-american/a-manuscript-47-years-in-the-making/, accessed October 2017.